# Landscape Reclamation - Rising From What's Left

*Edited by Luis Loures*

Published in London, United Kingdom

IntechOpen

*Supporting open minds since 2005*

Landscape Reclamation - Rising From What's Left
http://dx.doi.org/10.5772/intechopen.78424
Edited by Luis Loures

Contributors
Xiaobing Hu, Zhang Yalin, Jamilah Mohd Salim, Mohd Razali Sheriza, Fadzilah Majid Cooke, Maizatul Azwa Radzi, Aroloye O. Numbere, Jason Gordon, Neşe Yilmaz Bakir, Luis Loures



First published in London, United Kingdom, 2020 by IntechOpen
IntechOpen is the global imprint of INTECHOPEN LIMITED, registered in England and Wales, registration number: 11086078, 7th floor, 10 Lower Thames Street, London, EC3R 6AF, United Kingdom
Printed in Croatia

British Library Cataloguing-in-Publication Data
A catalogue record for this book is available from the British Library

Additional hard and PDF copies can be obtained from orders@intechopen.com

Landscape Reclamation - Rising From What's Left
Edited by Luis Loures
p. cm.
Print ISBN 978-1-78984-762-8
Online ISBN 978-1-78984-763-5
eBook (PDF) ISBN 978-1-83968-523-1

For EU product safety concerns:
IN TECH d.o.o., Prolaz Marije Krucifikse Kozulić 3, 51000 Rijeka, Croatia, info@intechopen.com or visit our website at intechopen.com.

# Meet the editor

Luis Loures is a landscape architect and agronomic engineer who holds a PhD in Planning and a Post-Doc in Agronomy. Since he graduated, he has published several peer reviewed papers at national and international levels and has been a guest researcher and lecturer both at Michigan State University (USA) and at the University of Toronto (Canada), where he developed part of his PhD research. During his academic career he taught several courses in different universities and countries mainly regarding the fields of landscape architecture, urban and environmental planning, and sustainability. Currently, he is a researcher both at VALORIZA and CinTurs, where he is involved in several financed research projects focusing on different investigation domains such as urban planning, landscape reclamation, urban redevelopment, and the use of urban planning as a tool for achieving sustainable development.

# Contents

# Preface

Landscape reclamation is generally considered a multidisciplinary field, which incorporates several branches of knowledge, integrating at the same time science, technology, and arts. For this reason, each and every specialist willing to reclaim derelict landscape should be perceived as someone who is able to promote the definition of multifunctional land uses, while enabling landscape alterations to ensure sustainable development, protect the environment, preserve natural and cultural assets, and improve people's quality of life.

In this regard it is increasingly important to reuse and rethink previously developed areas that are now abandoned or underused. Instead of consuming green lands, brown and gray lands need to be redeveloped and given new life, achieving a more sustainable urban setting. In this sense, although landscape reclamation plays a very important role in societal development, the continuous use of green fields continues to have deep economic, social, and ecological impacts that require special attention. The new environmental paradigms associated with globalization, progressive climate change, and increasing food production needs will certainly intensify the entropy and instability in most of the existing natural lands. This reality creates the perfect momentum to assess these issues.

By highlighting a body of knowledge related to the discussion of the opportunities and challenges associated with the development of new sustainable landscapes raised out of ruins, and considering current and future challenges related to landscape reclamation, planning, and development, the present book focuses not only on the different sustainability pillars, but also on the impacts these changes might have on each one of them.

Crossing a wide range of research domains and issues associated with derelict landscape redevelopment such as landscape reclamation, public involvement, landscape quality assessment, land use resilience, land policies, and urban planning, among others, this book intends to assess the impact of contemporary needs, constraints, landscape reclamation methods, and strategies on planning, ecosystem, and landscape design.

As a landscape architect and agronomic engineer with research interests deeply embedded in the fields of sustainability, postindustrial landscape planning, and design, my main investigation goals are directly connected to fitting design, especially that related to landscape transformation of previously developed or derelict areas, to the needs and desires of contemporary life, while addressing in equal measure society, the natural landscape, heritage and culture, and economic issues.

I truly believe that landscape reclamation processes are still missing the vision of circular planning, which differs from many contemporaries in its philosophical grounding in the social as well as creative matrices. This calls for a comprehensive view of the different components of landscape design, acknowledging the need for an interrelated analysis of the ecologic, cultural, and socioeconomic issues in landscape reclamation processes.

**Luis Loures**
Instituto Politécnico de Portalegre,
Portalegre, Portugal

Section 1

# Introduction

Chapter 1

# Introductory Chapter: Landscape Reclamation as a Key Factor for Sustainable Development

*Luis Loures*

## 1. Landscape reclamation: theoretical evolution vs. practical achievements: a brief overview

Landscape is continuously changing [1, 2] as a result of complex and interacting natural processes coupled with planned and unplanned actions by man [3]. This scenario of landscape transformation worldwide "*has raised global concerns*" ([4], p. 326), as it is the need to rethink landscape while protecting the environment. This is especially true for previously developed areas that are now abandoned or underused. Instead of consuming green lands, the brown lands need to be redeveloped and given new life, achieving a more sustainable urban setting [5–7].

In fact "*it has long been realized that urban planning and open space preservation are part of the same process*" ([8, 9], p. 273), "*and that the most effective way to protect open space is by effectively containing and managing urban growth*" ([8, 10], p. 273). In this regard, land transformation policies, strategies, methodologies and processes have been considered an important tool for urban containment, fostering urban redevelopment and revitalization [11–15].

Still, has shown by Loures [16] it is clear that these contributions and the principles they integrate, have not been adequately assessed regarding land transformation efforts. However, this approach may be considered a proficient method to address urban sprawl, increasingly viewed as significant and growing land-use problem that encompass a wide range of social, economic and environmental issues [8, 17, 18].

The relevance and popularity of landscape reclamation and landscape transformation approaches and projects are increasingly recognized and as referred by Reed [19] "*nearly every significant new landscape designed in recent years occupies a site that has been reinvented and reclaimed from obsolescence or degradation, as cities in postindustrial era remake and redefine their outdoor spaces*" (**Figures 1-4**).

Consequently, questions such as: What should be done with these landscapes? Which functions might these areas acquire in the future? What makes these spaces underutilized? What obstacles keep these landscapes from being transformed? Who is responsible for transforming them? Who is best qualified to do it? Is this process a single profession endeavor? Among others, remain to be answered. For this reason, new methodologies and frameworks are needed. In a period when "*(...) that seemingly old-fashioned term landscape has curiously come back to vogue*" ([23] in [24], p. 23), it is urgent to reinvent the way in which these derelict landscapes are transformed, considering not only environmental issues but also historic and cultural values, economic opportunities, and social needs.

**Figure 1.**
*Millennium Park, Chicago—view from Sears tower. Loures [20].*

**Figure 2.**
*Duisburg Nord Park, view from the Emscher river side. Loures [21, 22].*

The origin of this growing concern may be traced from a period when industry, became one of the main protagonists in the transformation of the city (Rossi, [25]). However, the consequences of the globalization of industry, relocation and restructuration of several industrial sectors over the past decades had a profound effect on quite a lot of industrial areas all over the world, producing a vast array of obsolete

**Figure 3.**
*Distillery District, Toronto, a multifunctional cultural asset from the city of Toronto. Loures [21, 22].*

**Figure 4.**
*Westergasfabriek, Amsterdam, channel side view. Loures [21, 22].*

industrial facilities with various impacts generated from them [26]. For this reason, numerous countries, all over the world, have undergone countless postindustrial land transformation projects (generally known as rehabilitation, revitalization, reclamation and/or redevelopment programs), in order to mitigate the negative effects arisen from these changes. In this scenario it is increasingly recognized that managing urban growth, transforming underused landscapes and protecting open space constitute relevant efforts to achieve sustainable urban planning.

Now no longer new, the production factories of the modern era have become obsolete, forcing this generation to decide on the disposition of the last generation's industrial environment. The international industrial climate, which Pirelli [27] has termed as the third industrial revolution has rendered obsolete several industrial structures, technologies and processes of the first half of the twentieth century. Demolition and abandonment were and continue to be "*fairly common approaches to deal with facilities that were designated as 'surplus' no longer serving their original production functions*" ([28], p. 48). Unfortunately, it is still common to find older buildings, characteristic of the industrial society, simply abandoned, surviving alongside with recent development areas. Nevertheless, the creation of new and more severe environmental legislation, and the public pressure related with the need to protect the environment, increased the necessity to redevelop derelict landscapes [29], considered by many as unrealized resources for initiating urban regeneration and ecological restoration [30–32]. Often in advantageous locations near city centers, along waterfronts, supported by existing infrastructure and adjacent to residential communities, these landscapes are environmentally impaired resources that need to be returned to productive uses, and reintegrated into the surrounding community [33]. Additionally, these land transformation projects, if developed at a larger scale and across multiple sites, could contribute to restore natural processes and functions, create multifunctional landscapes and promote sustainable growth [34].

## 2. Landscape reclamation: a multiplicity of activities towards sustainable development

The complexity inherent to the majority of current landscape reclamation projects, evident in the number of different ways in which they have been characterized, both in the literature and by designers and other specialists who worked and/or analyzed them, make derelict landscape redevelopment difficult to accomplish. Apart from eminent contamination and liability on many of these landscapes [35–37], redevelopment processes have to consider also planning, real estate transaction and land use issues [38–40], plus community and economic development issues [40–43], among others.

Considering this background and current need to reclaim derelict landscapes, this book will address both planning and design issues related to derelict land transformation. In fact, as mentioned by (Commoner [44], cited by [45]), thought the main problem lies in our means of production, in order to solve our derelict land problems, we need to change not only the location of certain activities but also the ways of making things. As it has been expressed, understanding this phenomenon is perhaps one of the most relevant consequences of assessing landscape reclamation issues, given that it becomes simpler not only to comprehend the current state of the art as it applies to us, but also to envision possible solutions for present and future problems [16, 21, 22, 29, 46–48].

As present trends of economic growth, resource consumption and environmental degradation become increasingly acknowledge as neither an acceptable nor sustainable option, discussion around why and how to redevelop derelict and or abandoned landscapes become progressively more relevant to growth management policies. As this remarkable phenomenon is gaining momentum, it becomes of utmost importance to address in one hand, the condition of these landscapes, and in the other the principles inherent to this process and the strategies and frameworks that best suit their redevelopment. For this reason, it is essential to study and understand both the differences between spaces generally typified as derelict landscapes, and the land transformation activities inherent to the redevelopment of these sites.

It is a given, that derelict land redevelopment, provides constant new opportunities for those who have the desire and the ability to seize landscape, regardless of their nature [6, 20, 49–51], for this reason landscape redevelopment and reclamation activities are considered to be, a significant resource for achieving sustainable development [52–55], contributing as well to improve life's quality. In this regard, reclamation processes need to be thought in terms of *sustainability* and/or *sustainable development*, terms that get used a lot these days, and which since their appearance have been faced as new development paradigms introduced in land-use matters, merging social, economic and environmental "dimensions" [56], and putting nations to work together in the definition of new principles and frameworks towards sustainable development.

Even if throughout recent years several normative theories associated to landscape reclamation, considering both design and planning principles towards sustainable communities, were created, the answer to this question is far from being achieved. From an overall viewpoint, sustainable landscape reclamation represents a subject of real sustainable dimensions, considering it is a positive response to environmental, social and economic issues [57, 58], which are the main dimensions of sustainability.

## Author details

Luis Loures
VALORIZA—Research Center for Endogenous Resource Valorization—Polytechnic Institute of Portalegre, Portugal

*Address all correspondence to: lcloures@ipportalegre.pt

## References

[1] Jinyan Z, Xiangzheng D, Tianxiang Y. Landscape change detection in Yulin prefecture. Journal of Geographical Sciences. 2003;**14**(1):47-55

[2] Pinto-Correia T, D'Abreu A, Oliveira R. Identificação de Unidades de Paisagem: Metodologia Aplicada a Portugal. In: Finisterra, XXXVI, 72. 2001. pp. 195-206. Available from: http://www.ceg.ul.pt/finisterra/numeros/2001-72/72_17.pdf [Retrieved: 12 August 2009]

[3] Loures L. Planning and design in post-industrial land transformation: East bank Arade river, Lagoa—Case study. Faculdade de Ciências e Tecnologia, Universidade do Algarve, Faro, Portugal—Dissertação de Doutoramento em Planeamento Urbano; 2011

[4] Musacchio L, Ozdenerol E, Bryant M, Evans T. Changing landscapes, changing disciplines: Seeking to understand interdisciplinarity in landscape ecological change research. Landscape and Urban Planning. 2005;**73**(4):326-338

[5] De Sousa C. Turning brownfields into green space in the City of Toronto. Landscape and Urban Planning. 2003;**62**:181-198

[6] Panagopoulos T, Loures L. Reclamation of derelict industrial land in Portugal: Greening is not enough. Book of Abstracts of the 10th European. Gelsenkirchen, Germany: Forum on Urban Forestry; 16-19 May 2007. pp. 71-72

[7] Portney K. Taking Sustainable Cities Seriously: Economic Development, the Environment, and Quality of Life in American Cities. Cambridge: MIT Press; 2003

[8] Bengston D, Fletcher J, Nelson K. Public policies for managing urban growth and protecting open space: Policy instruments and lessons learned in the United States. Landscape and Urban Planning. 2004;**69**(2-3): 271-286

[9] Hollis L, Fulton W. Open space protection: conservation meets growth management. 2002. Available from: http://www.brook.edu/dybdocroot/es/urban/publications/hollisfultonopenspace.htm [Retrieved: 26 April 2009]

[10] Alterman R. The challenge of farmland preservation: Lessons from a six-nation comparison. Journal of the American Planning Association. 1997;**63**(2):220-243

[11] Loures L, Burley J. Post-industrial land transformation—An approach to socio-cultural aspects as catalysts for urban redevelopment. In: Urban Planning. IntechOpen; 2011. pp. 223-246

[12] Loures L, Crawford P. Democracy in progress: Using public participation in post-industrial landscape (re)-development. WSEAS Transactions on Environment and Development. 2008;**4**(9):794-803

[13] Willem K. Taxing land for urban containment: Reflections on a Dutch debate. Land Use Policy. 2009;**26**(2):233-241

[14] Urban Land Institute, Barriers and Solutions to Land Assembly for Infill Development. Washington DC: The Urban Land Institute; 2004

[15] Adams D, Watkins C. Greenfields, Brownfields and Housing Development. Oxford: Blackwell Science Ltd; 2002

[16] Loures L, Burley J, e Panagopoulos T. Postindustrial landscape redevelopment: Addressing the past, envisioning the

future. International Journal of Energy and Environment. 2011;**5**(5):714-724

[17] Brueckner J. Urban sprawl: Diagnosis and remedies. International Regional Science Review. 2000;**23**(2):160-171

[18] Johnson M. Environmental impacts of urban sprawl: A survey of the literature and proposed research agenda. Environment and Planning A. 2001;**33**(4):717-735

[19] Reed P, editor. Groundwell: Constructing the Contemporary Landscape. New York: The Museum of Modern Art; 2005

[20] Loures L, Santos R, Panagopoulos T. Urban parks and sustainable city planning—The case of Portimão, Portugal. WSEAS Transactions on Environment and Development. 2007;**3**(10):171-180

[21] Loures L. Industrial heritage: A gear to redevelopment. In: Proceedings of the EURAU 08—Cultural Landscape, 4th European Symposium on Research in Architecture and Urban Design, 16-19 January 2008; Madrid, Spain; 2008a. pp. 1-7

[22] Loures L. Industrial heritage: The past in the future of the city. WSEAS Transactions on Environment and Development. 2008b;**4**(9):784-793

[23] Corner J. Terra Fluxos. In: Waldheim C, editor. The Landscape Urbanism Reader. New York: Princeton Architectural Press; 2006

[24] Waldheim C, editor. The Landscape Urbanism Reader. New York: Princeton Architectural Press; 2006

[25] Rossi A. The Architecture of the City. Cambridge: MIT Press; 1982

[26] Antrop M. Changing patterns in the urbanized countryside of Western Europe. Landscape Ecology. 2000;**15**:257-270

[27] Pirelli L. Progetto Bicocca. Milano: Edizioni Electa SpA; 1987

[28] Rea C. Rethinking the industrial landscape: The future of the Ford Rouge Complex [Master Thesis]. Cambridge: Massachusetts Institute of Technology; 1991

[29] Loures L, Horta D, Santos A, Panagopoulos T. Strategies to reclaim derelict industrial areas. WSEAS Transactions on Environment and Development. 2006;**2**(5):599-604

[30] Allen B, Linden M, editors. De-Industrialization: Social, Cultural and Political Aspects. Cambridge: University Press; 2002

[31] Backhaus G, Murungi J, editors. Transformation of Urban and Suburban Landscapes: Perspectives from Philosophy, Geography, and Architecture. New York: Lexington Books; 2002

[32] Brebbia A, Almorza D, Klapperich H. Brownfield Sites: Assessment, Rehabilitation and Development. Southampton: WIT Press; 2002

[33] Ekman E. Strategies for reclaiming urban postindustrial landscapes [Master thesis]. Massachusetts: Institute of Technology; 2004

[34] Collins T. Art and ecological restoration in cities. In: Hall T, Miles M, editors. Urban Futures. London: Routledge; 2001

[35] Alberini A, Longo A, Tonin S, Trombetta F, Turvani M. The role of liability, regulation and economic incentives in brownfield remediation and redevelopment: Evidence from surveys of developers. Regional Science and Urban Economics. 2005;**35**(4):327-351

[36] Gibbons J, Attoh-Okine N, Laha S. Brownfields redevelopment issues revisited. International Journal of Environment and Pollution. 1998;**10**(1):151-162

[37] McGrath T. Urban industrial land redevelopment and contamination risk. Journal of Urban Economics. 2000;**47**(3):414-442

[38] Amekudzi A. Integrating brownfields redevelopment with transportation planning. Journal of Urban Planning and Development. 2004;**130**(4):204-212

[39] De Sousa C. Brownfield redevelopment in Toronto: An examination of past trends and future prospects. Land Use Policy. 2002;**19**:297-309

[40] De Sousa C. Unearthing the benefits of brownfield to greenspace projects: An examination of project use and quality of life impacts. Local Environment. 2006;**11**(5):577-600

[41] Kaufman D, Cloutier N. The impact of small brownfields and greenspaces on residential property values. Journal of Real Estate Finance and Economics. 2006;**33**:19-30

[42] Ozdil T. Assessing the economic revitalization impact of urban design improvements: The Texas Main Street Program [Doctoral Dissertation]. Texas: Texas A&M University; 2006

[43] Paull E. The environmental and economic impacts of brownfields redevelopment. 2008. Available from: http://www.nemw.org/images/stories/documents/EnvironEconImpactsBFRedev.pdf [Retrieved: 15 March 2009]

[44] Commoner B. Making Peace with the Planet. New York: Pantheon Books; 1990

[45] Lyle J. Regenerative Design for Sustainable Development. New York: John Wiley and Sons; 1994

[46] Loures L, Panagopoulos T. From derelict industrial areas towards multifunctional landscapes and urban renaissance. WSEAS Transactions on Environment and Development. 2007a;**3**(10):181-188

[47] Loures L, Panagopoulos T. Sustainable reclamation of industrial areas in urban landscapes. In: Kungolas A, Brebbia C, Beriatos E, editors. Sustainable Development and Planning III. Southampton: WIT Press; 2007b. pp. 791-800

[48] Loures L, Panagoupolos T. Recovering derelict industrial landscapes in Portugal: Past interventions and future perspectives. In: The 3rd International Conference on Energy, Environment, Ecosystems and Sustainable Development; 24-26 de julho de 2007; Agios Nikolaos, Crete Island, Greece; 2007c

[49] Ferreira V, Panagopoulos T, Andrade R, Guerrero C, Loures L. Spatial variability of soil properties and soil erodibility in the Alqueva dam watershed, Portugal. Solid Earth. 2015;**7**(301-327):2015

[50] Loures L, Loures A, Nunes J, Panagopoulos T. Landscape valuation of environmental amenities throughout the application of direct and indirect methods. Sustainability. 2015;**7**(1):794-810

[51] Lu D, Burley J, Crawford P, Schutzki R, Loures L. Quantitative methods in environmental and visual quality mapping and assessment: A Muskegon, Michigan watershed case study with urban planning implications. In: Urban Planning. IntechOpen; 2011. pp. 127-142

[52] Castanho R, Loures L, Cabezas L, Fernández-Pozo L. Cross-border

cooperation (CBC) in Southern Europe—An Iberian case study. The Eurocity Elvas-Badajoz. Sustainability. 2017;**2017**(9):360

[53] Castanho R, Loures L, Fernandez J, Pozo L. Identifying critical factors for success in cross border cooperation (CBC) development projects. Habitat International. 2016;**72**:92-99

[54] Rato Nunes J, Ramos-Miras J, Lopez-Piñeiro A, Loures L, Gil C, Coelho J, et al. Concentrations of available heavy metals in Mediterranean agricultural soils: A case study in typical Mediterranean soil. Sustainability. 2014;**6**(12):91249138

[55] Vargues P, Loures L. Using geographic information systems in visual and aesthetic analysis: The case study of a golf course in Algarve. WSEAS Transactions on Environment and Development. 2008;**4**(9):774-783

[56] Singh P, Sharma A. Integrated approach to improve quality of life in urban distress areas sustainable urban regeneration. The International Journal of Environmental, Cultural, Economic, and Social Sustainability. 2009;**5**(2):121-128

[57] Loures L, Panagopoulos T, Burley J. Assessing user preferences on post-industrial redevelopment. Environment and Planning. B, Planning & Design. 2016;**43**(5):871-892

[58] Loures L, Vaz E. Exploring expert perception towards brownfield redevelopment benefits according to their typology. Habitat International. 2016;**72**:66-76

Section 2

# Landscape Reclamation, Processes, Techniques and Approaches

Chapter 2

# Project-Based Urban Renewal and Transformation of Urban Landscape in Turkey

*Neşe Yılmaz Bakır*


## Abstract

Nowadays, the effects of capital-based decisions and, therefore transformation has effects on the structure of the existing city, the physical, social, and economic future of the people living in that city, and consequently all the traditions of the city. In this process, the urban landscape rapidly changes and the public urban landscape areas are replaced by the private landscape areas in many cities. In the recent period, within the scope of space's meaning changing for the capitalism and capital accumulation, the cities are rapidly renewed with a model that can be named "project-based urban renewal." The objective of this research is to determine the problems of project-based urban renewal approach and to examine the effects on the urban landscape in Turkey. For this purpose, Kayseri city that comes to the forefront with its planned development history since the proclamation of the Republic in Turkey was selected as the study area. In the study, the project-based renewal projects built in Kayseri province and the changes in the urban landscape were comparatively examined at urban level and structure level in terms of uniform structuring, increase in the density, devastation in the green system, privatization of the public space, and gentrification.


**Keywords:** urban renewal, urban landscape, project-based renewal, planning systems

## 1. Introduction

The cities transform through the economic, demographic, social and ecological processes accompanied by the significant functional and structural changes in the urban landscapes. The urban landscape is constantly transforming into a different, because it has a sensitive structure that it carries the records and traces of events and movements [1]. Together with the residential, commercial, industrial, government-institutional, cultural-educational land uses, the patches of remnant vegetation, the secondary green areas such as parks and/or cemeteries, and the lands used for other purposes, the urban landscape mosaic constitutes is a quite complex structure [2].

In the literature, the concept of landscape refers to the complex transformation of the spatial structures through the social and cultural processes and it also refers to the interconnectedness between the spatial and the social aspects [3]. In the formation of urban landscapes, the data related with the socio-cultural structure is under the significant influence, as well as the physical structure.

From this aspect, the social, economic, environmental, natural and technological transformations also affect the urban area and, thus, the urban landscape constantly changes. The industrial revolution and also the population growth in the late nineteenth century accelerated the process of change. In the recent years, the economic developments but especially the neoliberalism placed the cities into the focus of the economy. The neoliberal policies are known to have spatial consequences [4]. As a coherent and long-term strategy arising from the neoliberal perspective, the production and consumption of the urban and metropolitan territories are debated. "Planning through the urban "projects" has been developed

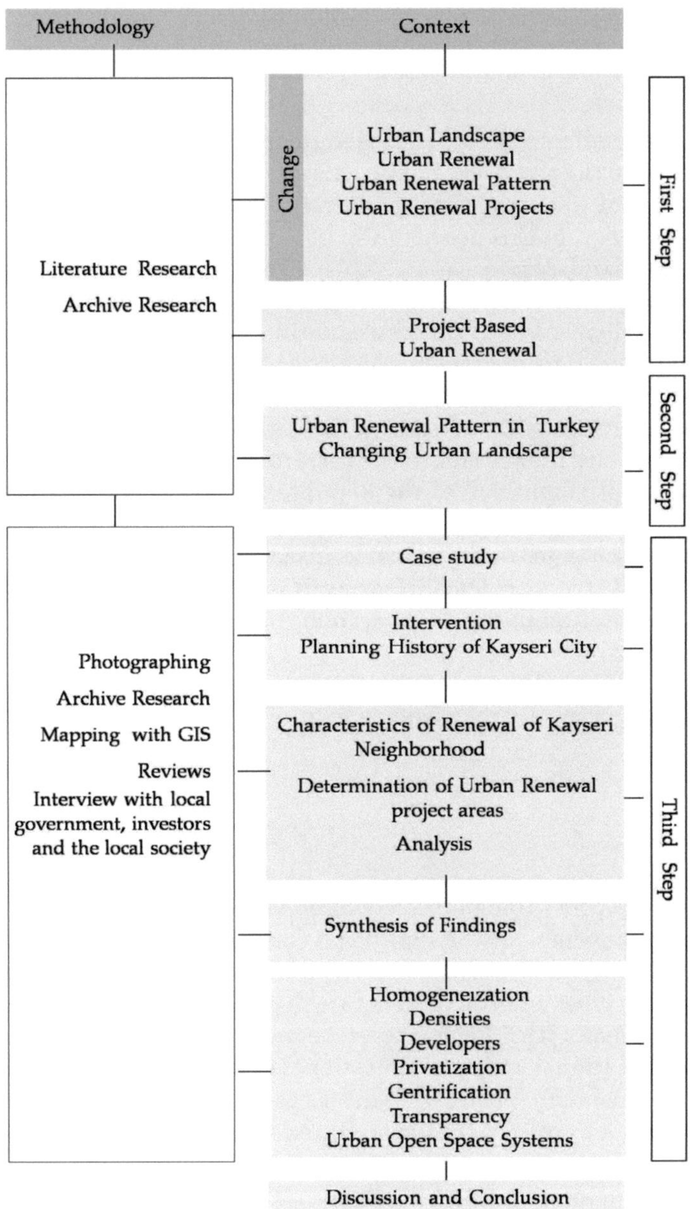


**Figure 1.**
*Methodological diagram of the study.*

as the main strategy of stimulating the economic growth" [5]. For this reason, the world witnessed the neoliberal strategies of the economy and governmental policies that started in North America and Western Europe since the late 1970s, then it has expanded elsewhere "in the hope of harmonizing (if not standardizing) economic and social policy" [6].

Together with the neoliberal transformation in the economy, "the development and expansion of industrial capitalism burgeoning the cities increasingly express the powerful impulse toward the centralization of capital" [7] that forces urban projects to re/organize the city space for the accumulation of labor and capital. As a result of this process, the poor and slum neighborhoods turned into the targets of profitable urban renewal projects.

In the present study, firstly, the change of urban renewal patterns and the project-based development process were examined theoretically by making use of literature research. The city of Kayseri, which is defined as the study area, has been developed in a planned manner since the proclamation of the Republic. The transformation in the urban landscape through the project-based urban renewal projects was analyzed together with the transformation planning processes by making use of the theoretical base developed. In this process, the detailed researches were carried out in the study area such as interviews and observations, archive researches, photo documentation. Also GIS-based mapping were used in revealing the overlapping between urban green areas and project areas in order to document the devastation in the green system (**Figure 1**).

## 2. The change of urban renewal pattern

The urban renewal projects of the nineteenth century focused on solving the problems of industrial cities, as well as sustaining the healthy and livable areas. The context of urban renewal projects dramatically changed especially after World War II. After the WW II, many cities of European countries faced problems such as economic decline, environmental destruction, and social dereliction. It is known that the State concentrates on the urban redevelopment, as well as the urban renewal, downtown revitalization, reconstruction of demolished areas, and public housing programs. Furthermore, the economic benefits of the urban redevelopment projects are also attention-grabbing for the investors and the State. Large amount of profit can be made by selling new and modern units at the city centers. The increase observed in the population density necessitates an increase in the employment opportunities. Moreover, it also accelerates the modernization process of city centers. Furthermore, it also became a tool for the state in order to prevent the inner cities from urban decline [8]. Thus, the main objective of the renewal projects in this period was to sustain the job opportunities by creating the flagship projects located at the older parts of the city centers, as well as increasing the demands for these areas.

The main focus of the urban renewal projects before the 1960s was on the eradication of the landscape observed after the WW II. Within the scope of these projects, the destructed and old houses were cleared off and the reconstruction process was initiated. The increase observed in the highway construction, on the other hand, created an increase in the number of cars owned. Because of these results, many residents left their homes. As a result of this process, many new problems emerged such as the relocation of residents. However, during this process, the commercialization of the city centers also increased and the number of people living in the residential areas decreased as a result of the displacement policies and suburbanization. Moreover, the importance given to protecting the cultural and natural heritage raised awareness about the historical locations at the city centers. After this Then, evaluating the existing large number of vacant housing stock in historical districts in inner cities became an important approach in the urban renewal agendas. Thus, the urban renewal performed during the aforementioned period can be called "urban rehabilitation" [9].

At the end of the 1970s, the inner city problems, especially depending on economic decline and environmental decay caused restless in societies. Depending on job losses and increases in the rents of houses, an increase was observed in the number of homeless people [10, 11]. The number of urban regeneration projects arose in the inner parts of the cities, where the economic decline, environmental decay, community dereliction, growing unemployment, and some social problems are observed [12]. The urban renewal projects aim to revitalize the old city centers. The common properties of urban renewal projects between the 1960s and 1970s were their comprehensiveness [13]. The urban renewal projects strategies refer to the social and physical rehabilitation of ancient areas in city centers.

Besides the effects of neoliberal political-economic factors, environmental factors too were important after the 1980s while implementing urban renewal projects strategies. According to Knox (1991), there are two important movements that have transformed the economic and socio-cultural structure of societies since the 1980s and they also influenced the urban renewal projects in the built environment of cities. The first among them was the transition from the Fordism (mass consumption and production) to the advanced capitalism concepts such as flexible accumulation, post-Fordism, and postmodernism. The second movement was based on "a philosophical, cultural and attitudinal differentiation from the modernism to the postmodernism" [14]. By making use of these concepts, Knox reported that the new urban patterns and landscapes are created by the relationships between the demand/consumption and the supply/production. The urban decline areas emerged in the built environment because of the changes observed in the demand and supply circuit. The changes in production also reconstructed the occupational structures. The advertising agencies, financial services, media specialists became new popular sectors for the last two decades. These sectors created a new bourgeoisie and also gained a place at the heart of the cities. The employees working at these industries moved from the suburbs to the city centers. The preservation of the old city quarters also attracts these groups and, thus, the gentrification became an inevitable consequence [14].

The tendencies for historic preservation, gentrification, or postmodern architecture became popular in the reformation of the built environment after the 1980s. These terms can be seen as the most visible reflections of the new policies of the new world order in the built environment. Similarly, the slums, old industrial quarters, and the old neighborhoods located at the city center were placed at the focus of the renewal projects of the neoliberal economy. Even though the urban renewal is not a new phenomenon that has emerged in the neoliberal era, the cities and rescaling projects became the key instruments for the entrepreneurial strategies aiming the economic success since the 1980s. The interests of the companies are determined in accordance with these strategies put into realization by the partnership between private and public sectors.

The urban renewal projects were the prominent urban strategy instruments in the 1990s, especially in the European countries, and these projects have been put into practice mainly for the revitalization, improvement, and preservation of the historical city centers or the industrial and commercial centers [15]. Keleş also stated that, over the last 20 years, the concept of regeneration turned from a physical definition into a more complex set of propositions that combine the social, cultural and economic objectives. In the majority of these regeneration projects, a significant amount of urban employment has been created [16].

The production methods and employment structure have also transformed and this transformation has also brought have new requirements from the aspect of urban systems. The traditional local economic progress became weaker because of the knowledge-based economy, negligence of the local interest, as well as the economic globalization. On the other hand, these factors also strengthened the effects of the external factors on the urban development. For instance, because of the globalization, the economic and cultural bonds in the city weakened and this caused the deepening social exclusion and deprivation (**Table 1**).

The cities of today promote themselves as a "world city," "global city," "knowledge city," "creative city," etc., in an entrepreneurial and competitive way. From the aspect of the competing cities, the neoliberal urban strategies such as privatization of public spaces, large-scale urban projects, residential housing projects, large-scale advertising and promotion campaigns of cities, highly speculative flagship or mega projects, dissemination of imaginary and brand cities, commodification of city centers, new consumerist practices of cities, and promotion of cosmopolitan city-center lifestyle come to the forefront [18–23].

### 2.1 Urban renewal projects

In literature, the urban renewal projects refer to a process of remodeling the urban areas by the means of rehabilitation, conservation, and redevelopment. The urban renewal projects are put into practice in various ways. The main urban renewal strategies can be exemplified as *urban revitalization, urban redevelopment, urban rehabilitation and urban regeneration* [24].

The *urban redevelopment* refers to destructing the existing buildings and changing the land use at that location [25]. However, the redevelopment approach also refers to the implementation of new projects replacing the existing building stock, which are in severely deteriorated status and have no preservation value or in which the arrangement of buildings cannot provide the satisfactory living conditions [8]. Moreover, this approach added new functional characteristics in order to revitalize the project area more from social and economic aspects. Generally, the redevelopments include the reconstruction of new buildings on the cleared land.

| Period | Urban renewal pattern |
|---|---|
| 1950 | The rebuilding of new urban uses for the replacement of old uses, the elimination of physical problems from the past and cities often based on a master plan, suburban growth |
| 1960 | Continuation of the 1950s approach. Suburban rehabilitate, peripheral growth |
| 1970 | Give priority to urban improvement and urban renewal projects<br>Acceptance of the link between physical deterioration and social distortion,<br>Projects focused on social problems |
| 1980 | Major schemes of development and redevelopment<br>Flagship projects downtown and outside<br>Cooperation with the public-private sector<br>Visible reflections of new policies of new world orders |
| 1990 | Move toward a more comprehensive form of policy and practice more emphasis on integrated treatments<br>Revitalization, improvement, and preservation of historical city centers or industrial and commercial centers |
| 2000 | The change in the concept of urban renewal from the "urban project"<br>Market-led and project-based development |

**Table 1.**
*The evaluation of urban renewal in Western countries [17].*

The urban redevelopment projects have been generally put into practice in the 1980s.

The *urban revitalization* refers to "the process, through which the disagreement between the services offered by the texture of the old regions and the contemporary needs can be eliminated" [26]. The urban revitalization aims to sustain the vibrant economy at the city center, as well as regaining the declining areas by developing new functions at those locations. The urban revitalization projects have played a dominant role in the declining areas since the 1960s. Doratli emphasized two types of urban revitalization [26]. The concepts stated by Doratli were the physical revitalization and the economic revitalization. The physical revitalization refers to the renewal incorporating the demolition, refurbishment, and conversion. The urban rehabilitation, on the other hand, means more than the revitalization since it incorporates the social improvement and the objective of raising awareness of society about the urban heritage. Moreover, the urban revitalization can be seen as an objective of urban rehabilitation processes [26]. The economic revitalization refers to a strategy that aims to create a vibrant economy in the old historical districts and economically disadvantaged regions. The historical urban regions play an important role for tourism and the economic potential of historical areas came to the forefront in last 30 years. The historical urban regions' adaptation to the global economic policies was continued by the strategies of urban renewal projects such as the urban revitalization. Moreover, the development of commercial and business centers in the declining segments of the city centers significantly contributed to the vibrant economic environments.

The *urban rehabilitation* refers to "the large-scaled interventions aiming to recover and update a lost or deteriorated function. The rehabilitation process includes different types of interventions ranging from the territory and urban areas to the building itself." The rehabilitation projects' main objective is to enhance the conditions of current building stock, infrastructure, as well as protecting the original character of the urban texture and removing the physical stock causing the urban decline [27]. The beginning of urban rehabilitation projects in the built environment in western countries dates back to the 1960s. According to Günay (1991), the urban revitalization and urban rehabilitation projects refer to the efforts aiming to keep existing inhabitants and property ownership pattern constant in the target area [27].

As stated by Weaver, there are two ways of applying or making use of the urban renewal. The first one refers to the interventions such as slum clearance and urban redevelopment. Constructing highways, establishing public works, and also the demolition and construction activities transform the physical structure of cities. The urban renewal projects fitting to this type have been widely put into practice between the industrial period and WW II [9]. The second use is urban regeneration programs for urban rehabilitation, which are financed by local and private funds. The second type of urban renewal projects has been widely applied, especially in the 1970s.

### 2.2 Project based urban renewal

The shifts in economies from the liberal to the neo-liberal policies increased the value of the city centers. The dominant approach of urban policies in the 1980s was the economic development based on the sustainability approach with the *project based urban renewal.*

In the recent years, the renewal is understood as a more controlled process that is carried out via different projects and strategies [28–30], whereas the regeneration (or revitalization) is explained as more spontaneous process taken place prevailingly in Central and Eastern European cities in the context of market-led urban development and slight intervention of public authorities [31–33]. Project-based projects, which are public-sector led and later partnership-based, are designed to capitalize on those sectors of economy which have growth potential [34].

The aforementioned process was a consequence of the market-based structure of the project-based urban renewal approach. In fact, the market-based urban renewal emerged after the political transformation and the emergence of market-based economies especially 1990. This segment refers to a large group consisting of a great variety of interventions with changing patterns of objectives, stakeholders, financial systems and political willingness, but characterized by some common elements, especially the extension of private financing and the relatively weakness of urban planning and regulation.

This fact does not exclude the presence of the programs incorporating the role of the public sector. The planned patterns of the urban renewal have two important common characteristics. First, all of the interventions were applied in parallel with the market-based interests and the main outcome that was expected was the increased attractiveness for the city or the neighborhood. So, gentrification was inevitably a result. Second, almost all of the interventions were limited to the physical renovation/renewal, whereas social, cultural, environmental, global urban factors remained secondary and they were often mostly ignored.

There was no doubt that these processes were differentiated every countries. For example, market-based urban regeneration in the Eastern-Central European countries was related to the transition of these countries and cities from a stat led system to a decentralized and market-oriented system. The most important of these processes was the reform of the housing system implemented in all countries, but there were only minor differences in character and timing.

As a consequence of the decentralization and privatization, the housing stock, which is a basic element of urban renewal, was depleted. The house ownership increased everywhere and public rental diminished. This development has a particular importance regarding urban renewal, and run-down inner city areas were affected in a specific way. While a growing number of inhabitants became owners of their housing, they often had no further capacities to contribute to the renewal of the common parts of the condominiums. As a result of the privatization, the local governments' physical intervention abilities were generally limited to the interventions in public areas, streets, squares, public buildings etc. However, they actually have no capacity to support the housing renovation. It can also be stated that this approach may be seen as the consequence of the *economic weakness of the local governments*.

In the literature, there are two methods of implementing the housing renewal; "property-led" and "area-based." The regeneration concept was developed in order to define the housing policies, legal tools, and programs aiming to re-organize the areas especially in the city center, which lost their functionality, that are transforming into areas of physical degradation and that need structural strengthening.

In the literature, the property-led regeneration practices refer to the prestigious projects having outstanding architectural and functional features, as well as the economic expectations they create [35]. The property-led regeneration significantly transforms the urban form on the specific decayed or deteriorated site and shares the same parcel units while neglecting social and economic sustainability at all. *Property-Led Development* can be defined as "the assembly of finance, land, building materials and labor to produce or improve buildings for occupation and investment purposes" [36]. The property-led regeneration involves the regeneration of an inner-city area by changing the image of the area, improving the environment, attracting private investment and improving confidence for further investment.

The main task of the public administration in property-led regenerations, which are led by the private sector, is described as the provision of a platform, coordination in capital stocks and investments, and efficient organization of local institutions. The task of coordination is suggested to bring together central government, related public institutions and local administration [37]. Despite the positive aspects of property-led regenerations from the aspect of economic development, the necessity of enhancing the role of inhabitants of deprived areas in the local economy is also criticized. On the other hand, the economic focus of the problem is frequently criticized because it causes an uncontrolled development [36].

*The area-based regeneration* refers to the redevelopment of a neighborhood that has integrity in its structure. The main objective here is to develop a program combining the physical, economic and social aspects of the physically degraded and economically disadvantaged neighborhoods. The property-led regeneration is an economy-based approach; however, the area-based regeneration concept initiated numerous programs and policies until the 2000s in many of the European countries particularly in France, the Netherlands, and the UK. The intention to bring physical, social and economic gain at the same time, the area-based regeneration programs have generally failed in achieving this goal [38].

In recent years, it was possible to identify a shift in the political discourse since the policy-makers and practitioners became interested in facilitating the involvement of local people in the process of developing the area-based regeneration initiatives because it is understood that the renewal begins from a proper understanding of communities. Furthermore, it is also clear that, even when the policy-makers and practitioners tried to employ a more bottom-up and community-centered approach in regeneration, the necessities on local administrations and other partners to incorporate the local society provided neither the time nor the resources to support this involvement. There are numerous different forms of the area-based policies. The most frequently seen type is a top-down mixture of different types of (physical, economic, social) interventions. Another type, which gained a place in the 2000s, was characterized with the efforts made in order to increase the role of local residents.

## 3. The change of urban renewal pattern in Turkey

The massive immigration from rural areas to big cities of Turkey began as a consequence of the industrialization. The increase in the population of metropolitan areas has reached high levels and the urbanization pressure has increased. Urbanization process started in a short time and the demand for urban land and housing increased to a very high level. As a result, especially housing needs of increasing population has been the most important agenda. These developments caused the urban renewal to the agenda and then the urban landscapes changed because of these practices (**Table 2**).

Turkey's urban developments in the 1980s and 1990s have remarkable similarities with the worldwide trends. Since the 1980s, the far-reaching administrative changes of the new era foreshadowed the upcoming urban developments. However, many trials have been failed until the 1990s. After that new strategies and approaches have been embraced in order to fix squatting and urbanization problems.

The increasing liberalization and globalization of the Turkish economy, further reflecting many other countries' experiences, the state in Turkey has played

| Periods | Urban renewal pattern | Changing urban landscape |
|---|---|---|
| 1923/1950 | Turkish cities to sustain physical transformation in the built environment under the effects of modernization movements | Grid urban texture, residential areas which made by Garden-city approach<br>*Build a new modern city* |
| 1950/1980 | Between the years of 1950 and 1980, the economic growth in large cities pulled people from rural to urban area.<br>As a result of the rapid urbanization, the vacant areas in large cities were transformed into squatter housing areas | Beginning in the 1950s and continuing throughout this period, the medium-rise (five to seven stories, two or three units per floor), reinforced-concrete frame apartment buildings on small urban lots became the generic residential typology in Turkish cities.<br>*Urban Apartments and Squatter Housing Areas* |
| 1980/2000 | Suburbanization, new residential developments were seen in the outer parts of the cities gentrification processes took place in historic districts<br>User-built first-generation *squatter* was progressively replaced by higher-rise, multi-unit apartments, now produced by a speculative process of commercialized, profit-driven, frequently illegal, and substandard construction. | The establishment of the Mass Housing Administration in 1984, with the aim of providing credit for large-scale production of low- and middle-income housing, has played an important role in the proliferation of standardized, multi-unit, high-rise blocks.<br>Higher rise (twelve to twenty-four stories and more) residential typologies, often built with prefabricated techniques and rationalized construction processes<br>*Residential towers, shopping malls, hypermarkets entire new edge cities* |
| 2000/today | Urban land became highly commoditized, the regulation of land market became an important asset of governments<br>• to prepare a legal basis for new investments for land development in urban development projects by privatization of state land<br>• announcement of urban transformation and development projects<br>• create new institutions<br>• (re) organize the market | Newly developed high-rise office buildings, luxury residences, gigantic shopping malls, mega-urban projects designed by star architects, numerous touristic entertainment facilities and mega-events, which are similar to urban projects elsewhere<br>Gated Communities and Suburban Expansion<br>*Project-based urban renewal* |

**Table 2.**
*Urban renewal and changing urban landscape in Turkey.*

a central role in engineering the market and the form of these urban projects. This nation-state led policy change had a significant consequence for the cities of Anatolia as they were able to nurture their export base and reach new international markets. At the same time, increasing decentralization efforts, again initiated by the nation-state, granted larger resources and greater responsibilities to the municipal governments. Thus, the local actors in the Anatolian cities gained increasing recourse to the market-based practices benefiting more from the market-based instruments of re-zoning urban land. The role of the State in the formation of the land and property market, its major role in designating and implementing the urban renewal projects, and its authority of determining the terms and conditions of the projects without allowing the participation of other parties define the current focus of urbanization as state-led urban development [21, 39].

The period of the 1980s constitutes a milestone in terms of changing the public response to the urban regeneration. The previous objectives (provision of housing, public amenities, and specific emphasis to the people having a low level of income) were put aside and the economic growth measured by the concentration of private investment became the sole criterion of the success for urban revitalization.

By the 2000s, the partnerships between local authorities and private sector arose. The urban renewal projects are implemented not only in slum areas but also in areas that are sensitive to natural hazards. Until 2012, there was no general law enacted for the urban renewal projects. The urban renewal projects were applied either region-specific laws or Law No. 5393 and 5272 municipality law. The urban renewal projects were applied by the authorized agencies. According to Article 73 of Law No. 5272 municipal law, these authorized agencies are metropolitan municipalities, country municipalities within the boundary of the metropolitan municipality, provincial municipalities, and municipalities larger than the population of 50,000. The current law applied to the urban renewal projects is Law No. 5393 municipal law.

According to this law, "The municipality shall implement urban transformation and development projects in order to create residential areas, industrial zones, commercial areas, techno-parks, recreational areas and any sort of social reinforcement areas, to reconstruct and restore the wearing segments of city, to protect the historical and cultural structure of city, and to take measures against the earthquake risk. In order for an area to be declared as urban transformation and development area, one or several of aforementioned criteria shall be found within the borders of municipal or neighboring borders." The law does not allow the metropolitan municipalities to declare urban renewal and development area without any limitation or supervision or allowance.

The 1999 İzmit earthquake was the crucial point for the urban renewal projects in Turkey. The government aimed to identify high-risk areas that are sensitive to possible natural hazards and re-arrange building stock that is out of standards. The government, private sector, and real estate investment trusts have attention on illegal and non-standard slums in city centers [40].

After 2011 van earthquake, the government took serious steps for demolishing illegal buildings and regenerating old ones; therefore Law No. 6306, known as "Urban Regeneration Law," officially named as "Law on Restructuring of Areas Under Risk of Natural Disasters" entered into force in May 2012.

After 2012, the urban renewal became one of the most frequently discussed problems in Turkey's urbanization process and practice. The law takes the earthquake risk as base and it addresses the renewal of buildings, which are in danger of an earthquake. The law defines the implementation processes and tools for both property- and area-based regeneration projects. Law No. 6306 introduced the term "risky building" and it defines this term as the buildings located in any area that is under the risk of an earthquake or the buildings scientifically and technically found to face the danger of collapse or get seriously damaged in an earthquake. The law simplifies the process of demolishing of a risky building and constructing a new one.

Since the entry of the Law No. 6306, a significant increase was observed in the numbers of property- and area-led regeneration projects. To date, due to their widespread effects on the socio-cultural and physical texture of the city and the project-based renewal projects were widely discussed. While the existing buildings are renewed through the property-led renewal, a significant transformation is also observed in the residential areas from the aspects of physical, social and economic environmental characteristics, as the overall effect of these implications.

The urban renewal projects are generally put into practice in order to improve the unplanned and problematic areas, occupied public lands, regions under disaster risks, and cultural and historic areas surrounded by the illegal settlements. It can be stated that the project areas are generally located around the city centers, where the land is of relatively higher value. In Turkey, the urban renewal projects were put into practice in order to convert the illegal squatter settlements into the well-planned modern commodities via the Housing Development Administration of Turkey and private corporations. Because, the squatter housing areas and old-historical quarters of cities do not only cause transformation in the physical structure of cities but they also affect the social, economic and environmental dynamics in the built environment. The municipalities with squatter or illegal established housing areas within their boundaries make use of the urban transformation and regeneration projects in order to enhance the living conditions and physical built environment with the standards of a contemporary lifestyle in transforming aforementioned areas into the prestigious regions [41].

In Turkey, it is understood that the "project-based method" is the most common and "market-based" and "area-based" methods for urban renewal projects. The Urban Renewal Project strategies such as urban rehabilitation, urban redevelopment, urban revitalization, and urban regeneration have an important place in the public discussions and the urban planning agenda, especially for the last 20 years. There are two main alternative approaches to implementing the model: (a) demolishing the illegal settlements, constructing new houses in the same area, and allocating them to the right holders, and (b) constructing new houses in a different area to transfer the right holders living in the upgrading area [42].

The investment in urban land and the formation of a speculative land market had an important effect on the urban economies and the development of new urban projects. The urbanization plan was designed on the basis of the idea of deindustrializing the metropolitan centers, which are intended to serve as the bases of the finance and service sectors. This laid the foundation of the state-led regeneration projects in the old industrial districts and working-class neighborhoods, as well as the gentrification of the neighborhoods located at the city centers and the megaprojects including gigantic shopping malls, high-rise office buildings, gated residential communities, and luxury condominiums.

As the urban land became highly commoditized, the regulations related with the land market became an important asset of governments in Turkey. The state has become one of the most important actors in the market, directing the privatization of state-owned lands, providing land for urban development projects, preparing legal grounds for new investments and announcing the fields of urban renewal and development projects [43].

## 4. Urban renewal and changing Kayseri city

The city of Kayseri, where the first practices of urban planning activities in Turkey were initiated and which remained at the forefront of the planned development process since the proclamation of the Republic, was selected as the study area. Kayseri's urban renewal process showed considerable similarities to the trends in Turkey. After the defeat of Ottoman Empire, the Turkish Republic was founded in 1923. In this period, the main purpose of the state was to reconstruct the national economy and make institutional developments in the economy. In order to create the new and modern environments, the state made reforms and applied new master plans for eliminating the effects of World War I and transforming the traditional Turkish society into a modern one. The national economic policies were applied in order to establish a bourgeoisie class and to fasten the social-economic transformation [44].

Big public works and urban reconstruction projects were applied in Turkish cities in the process of the establishment of the new country. Danger plan for Izmir, prost plan for Istanbul and Öelsner-Aru plan for Kayseri are examples of master plans at this period [44, 45]. The planned development and modernization activities played important role in Kayseri since the proclamation of the republic. Until the 2000s, it can be said that the renewal processes of the city have developed according to the plan (**Table 3**).

This city was partially planned for the public investments in the 1/8000 scale schematic Çaylak plan, which has been prepared in 1933. As a reflection of this modernist approach, the urban plan of this city was prepared by Kemal Ahmet Aru-Öelsner from a holistic approach in the year 1945. In the plan prepared by him, the gridded urban design incorporating the wide boulevards, which was accepted as the main representation of those years and constituting the main pattern of the city, became dominant.

After the 1950s, the squatter housing areas became apparent in city pattern. The migration from rural to urban areas formed housing problem in cities, because the housing stock was not enough for newcomers. They constructed substandard housing units on the public land. At the beginning of the 1960s, some of the squatter housing districts transformed into illegal, and high rise apartment stocks, whereas the vote potential of squatter housing districts has been used by politicians. In this period, Kayseri began to canalize their capital, which they had been accumulating through commerce, into industrial investment and, as a result, urban mobilization began.

Another important period is the one after 1970. The second master plan prepared by Yavuz Taşçı, an architect, in 1975, the city center was reinforced and a

| Plans | Impact of plan on urban landscape | Urban renewal patterns |
|---|---|---|
| 33 Çaylak plan | Modernist<br>The distinction of the old and new city part<br>Grid system neighborhood<br>Establishment of public spaces | Urban rehabilitation (on existing urban texture) |
| 1945 Oelsner-Aru plan | Reflection of modernization on space<br>Destruction of an urban site<br>Grid system<br>Garden city effect<br>Detached, extroverted houses | Urban revitalization (on existing housing stock)<br>Clearance (on traditional housing texture) |
| 1975 Taşçı plan | Linear development<br>Sectoral distinction in the city<br>Metropolitan city vision<br>Identify urban sites<br>Transition to multistory construction | Urban rehabilitation (on existing urban texture)<br>Urban conservation (on urban historical site) |
| 1986 Topaloğlu-Berksan plan | Linear development<br>Metropolitan city approach<br>Shrinkage of historical site boundaries<br>Multistory construction<br>Building layout flexible/uncertain | Urban rehabilitation (on existing urban texture)<br>Urban redevelopment<br>Urban renewal (on squatter areas) |
| 2006 Doğan plan | Urban expansion as a spreading "oil stain"<br>Radial urban texture<br>Building pressure on natural and urban sites<br>Density increase<br>Transformation of squatter areas into multistory residential areas<br>Gated sites<br>Mass housing construction<br>Expansion of municipal boundaries and merger with semi-rural settlements<br>Grid urban texture<br>Urban projects<br>Urban renewal as an intervention tool | Urban revitalization (on existing housing stock)<br>Urban rehabilitation (on existing urban texture)<br>Urban conservation (on urban historical site)<br>Urban renewal (on squatter areas)<br>Urban redevelopment (Gated sites) |

**Table 3.**
*Urban landscape and urban renewal pattern in Kayseri.*

development plan with a single center and linear form were designed. In this period, the city's traditional districts (which were filled with ornate mansion-houses mostly dating from the eighteenth and nineteenth centuries) were demolished. Besides, the historical trade center was transformed into the modern city business center. New housing development areas were set across from the west to the east side of the city. Therefore, the city was converted from compact to linear form. The development plan suggested constructing broad boulevards and high-rise buildings which are still characteristic features of the city of Kayseri [45–47].

In 1986, a new master development plan was prepared by Topaloğlu and Berksan. Similar to the previous period, high-rise buildings and housing projects for middle and upper-income groups increased and the use of new materials and techniques was common in this period [45]. The first renewal projects were performed at the parcel-level by contractors.

During the 1990s, in Kayseri, market-based policies became more important for the city. The period that began under Karatepe's leadership in 1994 demonstrated that these municipal practices favored export-oriented policies and the liberalization of the Turkish economy [48, 49]. The privatization of municipal services continued at a greater level, especially after 1999. This process entailed a model popularly known as build-operate-transfer. During this period, the basic spatial practices were large-scale housing projects, transport and infrastructure projects, thematic-parks and sports facilities, museums, historic urban texture renewal projects, traditional public spaces projects (streets, squares, and parks) and mixed-use projects [45, 48].

Depending on the initiatives taken in laws regarding the urban transformation in the 2000s, the renewal practices gained significant speed in Kayseri city and entire country world In Kocasinan district, Ziyagökalp, Yenidoğan, Seyrani, Ahi Evran, Yunusemre, Argıncık, Yeşil Mahalle, Kuşcu, Oruçreis, Mithatpaşa, Erkilet, Yıldızevler, and Uğurevler neighborhoods were announced as the urban renewal areas under the conditions specified by law on municipalities No. 5393 Art. 73. But, besides these squatter areas, also the neighborhoods that are nearby the city center

| District | Neighborhood | Area (ha) | Population (persons) |
|---|---|---|---|
| Kocasinan | Ahievran | 25.5 | 240 |
| Kocasinan | Cırkalan | 260.55 | 397 |
| Kocasinan | Sahabiye | 50 | 5341 |
| Kocasinan | Seyrani | 5.1 | 340 |
| Kocasinan | Uğurevler | 87.23 | 6240 |
| Kocasinan | Yunus Emre | 7.3 | 780 |
| Kocasinan | Yıldızevler | 27 | 1628 |
| Kocasinan | Ziya Gökalp | 23.25 | 1568 |
| Melikgazi | Anbar | 5.8 | 268 |
| Melikgazi | Karacaoğlu | 3.7 | 284 |
| Melikgazi | Küçük Ali | 3.5 | 336 |
| Melikgazi | Kazım Karabekir | 32 | 1392 |
| Melikgazi | Yeni Mahalle | 85.53 | 2752 |
| | Total | 593.21 | 21,566 |

**Table 4.**
*Urban renewal areas in Kayseri province.*

and have a historical background and subjected to habitability certification such as Sahabiye, Küçük Ali, Battalgazi, and Karacaoğlu neighborhoods were also declared as the urban renewal zones (**Table 4**).

In Melikgazi district, the project processes of Kazım Karabekir and Anbar neighborhoods within the scope of law No. 6306 on renewal of regions under the risk of disaster were almost completed, and the implementation stage was started. Besides the renewal of 593.21 ha area influencing 21,000 users in Kayseri province, also the urban renewal requests were placed for Argıncık, Yeşil Mahalle, Kuşçu Mahallesi, Mithatpaşa, Erkilet, Bahçeşehir, Yavuzlar, Oruç Reis, Pervane neighborhoods and (in Melikgazi district) Kılıçarslan, Battalgazi and Seyitgazi neighborhoods.

## 5. Findings and discussion

Thirteen areas, which are to be subjects of a renewal project, have been determined in the light of the information gained from Melikgazi and Kocasinan. According to the examinations, the formation ideas of the areas in Kayseri, which were 1944 Oelsner-Aru plan, 1975 Yavuz Taşçı plan, 1986 Toplağlu- Berksan, 2006 Doğan plan already transformed or whose renewal decisions were taken, are mostly based upon 2006 plan decisions. The reason for this situation, after 2000, as a result of the economic growth of the city of Kayseri, it is the increase of renewal pressure in the space depending on the competitive processes.

Kayseri is going through a period of urban projects changing the urban landscape. However, the previous researches showed that the number of holistic projects with economic, social, and physical objectives is limited. Investigating Kayseri from the aspects of the basic characteristics of urban regeneration areas and categorizing the city as area-based and site-specific projects, it can be seen that the site-specific projects are at the forefront.

When we consider the renewal process as a whole; we can identify the things below;

### 5.1 Homogeneous architecture and urban environments

Even handle with best intentions and professional care, larger areas designed by single designers or groups of designers lack the heterogeneity of urban neighborhoods (**Figure 2**).

The Turkish word for "neighborhood" is mahalle. While the mahalle is the urban residential space, this word also refers to a space of social memory in Turkish popular culture defined by familiarity, belonging and tolerance in a local. In these areas, where people feeling belonging to the neighborhood live, people are in a close relationship in their daily lives. The physical renewal affects the social structure deeply (**Figure 3**).

In the neighborhoods that have been partially transformed, there are problems of not only of the new gentry but also a homogenous urban usage. These areas

**Figure 2.**
*Homogeneous architecture (Karacaoğlu, Anbar, Mithatpaşa Neighborhoods, 2018) [49].*

quickly turned into the centers of new life and got invaded by the cafes, restaurants, boutiques. For this reason, they became the areas attracting tourists at most and losing the sense of a real urban neighborhood.

### 5.2 Possible gentrification

Without careful public interference, private initiatives as well as government-led projects both lead to a complete change of inhabitants as well as urban character. Before the projects, the social geography of the neighborhoods was marked by its heterogeneous population, which suffered from poverty and the impacts of forced migration to this area because of poverty concentrated in this neighborhood. The central location of the neighborhood, which offers easy access to the informal labor markets in the center, cheap rent levels available abandoned building stock underlies the existence of very diverse and the least privileged groups in the neighborhood. After this project, the people living in the area started to leave there. No social policies, programs were integrated into the culture and tourism-based urban renewal scheme and no fixed measures were undertaken in order to keep the current population of Kayseri urban site in the area while improving their living conditions.

### 5.3 Extreme densities

In the case of renewal, in accordance with the general view, urban densities need to be increased in order to create a viable economic model without re-placing original inhabitants. However, in areas such as Sahabiye, Yeni Mahalle and Ahi Evran neighborhoods (**Figure 4**), higher densities increase the pressure on infrastructure, roads and public facilities. Because of these insufficient systems, quality of life is reduced and does not provide planning standards. In addition, these projects are exemplary and rapid transformations are observed in the neighborhoods, city centers and city boundaries after these applications.

A new texture, that cannot be adapted and articulated with the existing texture is formed, the urban landscape is deteriorating in its entirety.

### 5.4 No interest by developers

In the case of models such as "Support," it may sometimes be very hard to draw the attention of developers because the areas in urgent need of transformation are not the primary targets in order to achieve a higher economic gain in the city.

Investors do not show any interest in places where weathered and unqualified housing stock, such as the Fevzi Çakmak Quarter and Argıncık Quarter, are high. The most important reason for this is the fact that because of the lack of

**Figure 3.**
*Homogenous urban environments (Kazım Karabekir Neighborhoods) (2018) [50].*

**Figure 4.**
*Extreme densities (Sahabiye Neighborhood) (2018) [51].*

attractiveness, investors believe that will not be able to get the economic value of their investments.

### 5.5 Transparency

Due to the controversial structures of renewal projects, decision makers and practitioners do not tend to be transparent in order to reduce public reaction and are trying to implement them in a hastily. The developers and designers of these projects are usually government-dependent and have the power to advance the system rapidly. The interviews conducted in the fields, usually until the last moment, it was seen that the people do not have any information that their neighborhoods have been declared as renewal area.

It was understood that the people living in the neighborhood obtained the information about the project from the visuals held in the negotiation process after the implementation decision and from the visuals displayed around the project area.

### 5.6 The regions have seen a mission developer

These areas have become an attraction for new investments. This entrepreneurial role is significant in the sense that the situation of severe disinvestment in the neighborhood would not be halted by private investors.

### 5.7 The deterioration of urban open space system

These projects, which take place on the urban green system, harm the integrity of the system after the transformation in the area (**Figure 5**). These areas, which mostly serve the integrity of gardens and parks and open, semi-open green space, make the system inadequate with new typology.

Another aspect of these constructed spaces is the way, in which the open spaces are used within these projects. Increasingly, the shopping malls and the other facilities are constructed with the functional open spaces around the buildings. Some of these are in the form of parks landscaped with exotic plants and benches, some others are designed as small plazas with marble pavements. In the beginning, these open spaces were just a part of the architecture of the projects. However, in the course of time, they turned into extensions of shopping malls and recreational facilities. Increasingly, the previously mentioned claim of providing a desirable public space to the residents and customers of the projects has evolved into a new form through the "public events" organized by the managements of these newly constructed projects.

**Figure 5.**
*Kayseri urban green areas [45] and urban renewal areas map.*

### 5.8 Privatization of public open spaces

These places they use are safe, monitored, clean and convenient in order to meet their demands, as well as free from the unwanted urban crowds. The major similarity between these projects is their claim to bring the public and private realms together. The perceptions of urban residents about the "publicness" of the urban space also have an important part in the formation of these projects. The ways in which the urban middle classes think about different parts of the city elevates the demand for the lifestyle offered by such projects and thereby encourage the investment to the construction of privatized public spaces.

## 6. Conclusion

The urban landscape is a complex structure which is a result of the interaction between human and his environment. It also involves a social dimension, a cultural dimension, and an economic dimension. Therefore, they inherit communities' values, beliefs, and symbolic meanings occurring and changing in the course of the time. They change as the communities change, the lifestyles change, and the global economy changes. The urban renewal or revitalization is related with not only the old or the other buildings and the environment or rent but also with the collective memory, symbolic meanings embedded to buildings and structures, and the values attached to the space for centuries by the people living in the area.

In Turkish metropolitan cities, similarly to Kayseri city, the most significant renewal projects of the past decade have been increasingly in the form of "project based" in terms of their size and cost. Urban renewal projects' main aims are to improve the social mix, to improve the environment, as well as the quality of life of inhabitants and city dwellers, to promote rehabilitation of complex urban structures, to preserve the valuable and unique fabric of the selected areas, to control-reverse the deterioration of specific urban zones (i.e. residential, commercial, etc.), to restructure economic activities located in the urban fabric, to sustain urban landscape. These are a very important issue in terms of the renewal process. Social goals may be the main focus in the context of a national policy for combining urban social fabric and larger social structure. Although sustainable development goals such as improved living conditions, existing structural equipment and risk prevention and protection are never ignored, they are rarely the main focus. Targets should also include:

- To restore buildings;
- Improvement of infrastructure and public systems;
- Identify, analyze and identify working needs and opportunities for selected areas (urban core, out-of-city, etc.);
- Developing new business and professional opportunities;
- Establishment of indirect measures (infrastructure) and direct instruments to promote economic growth;
- Organize capacity building for institutions and organizations responsible for urban management and physical planning;
- Strengthening the provision of community-based integrated core services for vulnerable groups;
- Strengthen communities by increasing their problem solving, management and negotiation skills;
- Promote public participation at the national level for urban renewal policy and support;
- Developing transparent and participatory policy.

However, as in many countries and in our country, there can be discrepancies in the theory and post-practice. The economic development should also be encouraged, especially in an era where the world order is designed by competitiveness; however, it should not be the primary goal and the considerations should not be separated from its role as a public entity. In addition, historical urban housing areas are no longer seen as "common public assets" and thus designated renewal areas are not viewed as society's common cultural capital.

## Author details

Neşe Yılmaz Bakır
City and Regional Planning Department, Erciyes University, Kayseri, Turkey

*Address all correspondence to: nyilmaz@erciyes.edu.tr

## References

[1] Loures L, Panagopoulos T. Sustainable reclamation of industrial areas in urban landscapes. WIT Transactions on Ecology and the Enironment. 2007;791-800

[2] Andersson E. Urban landscapes and sustainable cities. Ecology and Society. 2006;**11**(1):34. [online] http://www. ecologyandsociety.org/vol11/ iss1/art34/

[3] Pakarinen T. Changing Landscapes. Nordisk Arkitekturforskning, Nr 2/1993. 2015:97-107

[4] Sager T. Neo-liberal urban planning policies: A literature survey 1990-2010. Progress in Planning. 2011;**76**(4):147-199

[5] Swyngedouw E, Moulaert F, Rodriguez A. Neoliberal urbanization in Europe: Large–scale urban development projects and the new urban policy. Antipode. 2002;**34**(3):542-577

[6] Jessop B. The Future of the Capitalist State. Cambridge: Polity; 2002. p. 216

[7] Smith N. New globalism, new urbanism: Gentrification as global urban strategy. Antipode. 2002;**34**(3): 427-450

[8] Broudehoux AM. Neighborhood regeneration in Beijing: An overview of projects implemented in the inner city since 1990 (doctoral dissertation). McGill University Libraries; 1994

[9] Weaver RC. Major factors in urban planning. The Urban Condition. L. J. Duhl, ed. New York: Basic Books; 1963

[10] Soja EW. Postmetropolis: Critical Studies of Cities and Regions. Oxford: Basil Blackwell; 2000

[11] Hall P. Cities of Tomorrow: An Intellectual History of Urban Planning and Design in the Twentieth Century. London: Routledge; 2002

[12] Jing-Xiang LJBZ. The urban renewal evolvement since modern times and thought of nowadays China urban renewal. Urban Problems. 2003;**5**:015

[13] Akkar ZM. Kentsel dönüşüm üzerine Batı'daki kavramlar, tanımlar, süreçler ve Türkiye. Planlama. 2006;**2**:29-38

[14] Knox PL. The restless urban landscape: Economic and sociocultural change and the transformation of metropolitan Washington, DC. Annals of the Association of American Geographers. 1991;**81**(2):181-209

[15] Roberts P. The evolution, definition and purpose of urban regeneration, in: P. ROBERTS and H. SYKES (Eds) Urban Regeneration: A Handbook, London: Sage. 2000:9-36

[16] Keleş R. Urban regeneration in Istanbul. In: Draft Paper to be Presented to Priority Action Program, Regional Activity Center. 2003. pp. 1-30

[17] Carter A, Roberts P. Strategy and partnership in urban regeneration. In: Roberts P, Sykes H, Grager R, editors. Urban Regeneration. 2nd ed. 2000. pp. 44-69

[18] Hubbard P. Hall T. The entrepreneurial city and the new urban politics, in Hall, T. And Hubbard, P. The Entrepreneurial City: Geographies of Politics, Regime and Representation, John Wiley & Sons, Chichester, UK; 1998

[19] MacLeod G. From urban entrepreneurialism to a "revanchist city"? On the spatial injustices of Glasgow's renaissance. Antipode. 2002;**34**(3):602-624

[20] Penpecioğlu M. Kapitalist kentleşme dinamiklerinin Türkiye'deki son 10

yılı: Yapılı çevre üretimi, devlet ve büyük ölçekli kentsel projeler. Birikim. 2011;**270**:62-73

[21] Şengül T. Kentsel Çelişki ve Siyaset. Ankara: İmge Yayıncılık; 2009

[22] Bakır NY. Kentsel Proje Kavramı ve Gelişim Süreci. Avrasya Terim Dergisi. 2013;**1**(2):74-82

[23] Zukin S. David Harvey on cities. In N. Castree and D. Gregory, editors, David Harvey: A Critical Reader. Oxford, Blackwell, UK. 2006:102-120

[24] Erbas AE, Erbil T. Neoliberal challenges and practices of urban regeneration projects in Istanbul. In: Real Corp Conference, "Planning Times, You Better Keep Planning or You Get In Deep Water, For The Cities They are a-Changin". 2013. pp. 843-852

[25] Keleş R. Kentleşme Politikası, Genişletilmiş 8. Baskı. Ankara: İmge Kitabevi Yayını; 2004

[26] Doratli N. Revitalizing historic urban quarters: A model for determining the most relevant strategic approach. European Planning Studies. 2005;**13**(5):749-772

[27] Duzcu S. The assessment criteria of urban regeneration projects: The case of the Fener-Balat districts in Istanbul (doctoral dissertation). Ankara: Middle East Technical University; 2006

[28] Ganser R, Williams K. Brownfield development: Are we using the right targets? Evidence from England and Germany. European Planning Studies. 2007;**15**(5):603-622

[29] Couch C, Petschel-Held G, Leontidou L, editors. Urban Sprawl in Europe: Landscape, Land-Use Change and Policy. John Wiley & Sons; 2008

[30] Tallon A. Urban Regeneration in the UK. Routledge; London, Routledge. 2013

[31] Feldman M. Urban waterfront regeneration and local governance in Tallinn. Europe-Asia Studies. 2000;**52**(5):829-850

[32] Brade I, Smigiel C, Kovács Z. Suburban residential development in post-socialist urban regions: The case of Moscow, Sofia, and Budapest. In: German Annual of Spatial Research and Policy. Berlin, Heidelberg: Springer; 2009. pp. 79-104

[33] Scott JW, Kühn M. Urban change and urban development strategies in central East Europe: A selective assessment of events since 1989. European Planning Studies. 2012;**20**(7):1093-1109

[34] Watson V. 'The planned city sweeps the poor away ...': Urban planning and 21st century urbanisation. Progress in Planning. 2009;**72**(3):151-193

[35] Loftman P, Nevin B. Prestige projects and urban regeneration in the 1980s and 1990s: A review of benefits and limitations. Planning Practice and Research. 1995;**10**(3-4):299-316

[36] Turok I. Property-led urban regeneration: Panacea or placebo? Environment and Planning A. 1992;**24**(3):361-379

[37] Imrie R, Thomas H. The limits of property-led regeneration. Environment and Planning. C, Government & Policy. 1993;**11**(1):87-102

[38] Aalbers MB, Van Beckhoven E. The integrated approach in neighbourhood renewal: More than just a philosophy? Tijdschrift Voor Economische en Sociale Geografie. 2010;**101**(4):449-461

[39] Keyder Ç. Istanbul into the Twenty-First Century. In: Orienting Istanbul, Cultural Capital of Europe?, Edited by Deniz Göktürk, Levent Soysal, Ipek Tureli, Routledge, London and New York; 2010:41-50

[40] Uzun CN. Yeni Yasal Düzenlemeler ve Kentsel Dönüşüme Etkileri. Planlama. 2006;**2**:49-52

[41] Ulu A. Regeneration of urban areas in cities of Turkey. Case study: Izmit. In: Conference on Sustainable Urban Areas, Rotterdam. 2007. pp. 25-28

[42] Mutlu E. Criteria for a "good" urban renewal project: The case of Kadifekale Urban Renewal Project (İzmir, Turkey) (master's thesis). İzmir Institute of Technology; 2009

[43] Ünsal BÖ, Türkün A. Neoliberal Kentsel Dönüşüm: Kentsel Alanlarda Sınıfsal Tahliye, Yoksullaşma ve Mülksüzleşme, Türkün, Asuman (Der.); Mülk, Mahal, İnsan; 2014

[44] Aksoylu S. A critical outlook to the planning practices of Turkey from the beginning of republican period. In: Hawaii International Conference on Social Sciences. 2003

[45] Yılmaz Bakır N. Kentsel Planlama Ve Proje Bütünleşme Süreci Kayseri Kenti Örneği (Yayınlanmamamış Doktora tezi). Mimar Sinan Güzel Sanatlar Üniversitesi Fen Bilimleri Enstitüsü Şehircilik; 2012

[46] Kocatürk F. Kayseri'de Konut Yerleşimlerinde Arz Talep değişimi (Doktora Tezi); 2004

[47] Doğan AE. Eğreti kamusallık: Kayseri örneğinde islamcı belediyecilik. Vol. 206; 2007

[48] Güner HE. Liseden Müzeye, Eğitimden Tüketime; Kayseri Lisesi'nin Milli Mücadele Müzesi'ne Dönüşümü. Kültür Envanteri. TÜBA-KED. 2017;**16**:67-87

[49] Available from: http://www.melikgazi.bel.tr/Kentsel-Donusum

[50] Available from: http://www.kocasinan.bel.tr/2018

[51] Available from: http://www.sahabiyedonusum.com/

Chapter 3

# Making Art Involve in the Paradigm of Ecological Landscape Construction in Ancient Villages—Taking Cai Fan and Longtan Ancient Village in Taihu as an Example

*Hu Xiaobing and Zhang Yalin*


## Abstract

The ancient residences of Cai fan Ancient Village and Longtan Village in Taihu County of Anqing City are historical and traditional villages with common regional cultural background. They have strong cultural characteristics of Huizhou and foreign architectural features. The overall architectural style of the ancient villages in Anqing area is combined with the topography, landforms and landscapes of the area where it is located, forming a unique architectural style. In addition, during the Ming and Qing Dynasties, the strong economic support of Huizhou merchants and the cultural education became increasingly prosperous. Developed, and even after the Huizhou merchants returned to their hometowns, they conceived and built houses with elegant, literary and detached mentality, which made the cultural environment and ecological landscape of Anqing ancient villages richer and the village landscape more prominent. The ancient village of Anqing is a typical representative of the regional culture of the southwest and has important historical and cultural values.


**Keywords:** ancient village, ecological landscape, text paradigm, Cai fan, Longtanzhai

## 1. Introduction

Throughout the history of the development of the world landscape, both the Western and Eastern landscape designs have been focused on architectural design, and the architectural structures, water bodies, plants, and mountain combinations in landscape science can form the elements of ecological beauty. As far as architecture is concerned, it is only one of the important elements in artificial landscape design. Architecture is the main carrier of people's life. It is precisely because of this modern construction, people's emphasis on nature, humanities, customs, and ecological landscapes has been weakened. People's increasing material and spiritual cultural needs and the need to optimize their living space have made landscape design a more urgent need for perfection. It has also become a weather vane to measure the quality of people's lives. Landscape design is not a narrow concept equivalent to general greening. It is a combination of urban planning, urban function, landscaping, architectural modeling, sculpture, installation art, environmental system, urban water supply, and drainage, a systematic project in which botany, esthetics and nature are integrated into harmony with nature. At present, the landscape design of beautiful villages is mostly a hard-won set of modern landscape design. It lacks understanding and respect for nature, humanities, customs, ecology, and characteristics [1]. It does not have all the characteristics, culture, and ecology of the original village. The rapid development of modern social economy has promoted the rapid development of material civilization. At the same time, people's demand for spiritual civilization is getting higher and higher. It is realized that the importance of ecological landscape makes people no longer simply obey the single line of architecture, but give much more attention to the two-line development direction of the building to match the ecologically harmonious landscape. Under such a background, the harmonious landscape of architecture and ecology will eventually become a two-line harmonious development direction and that will become the major impact of production of human, living space and quality of life, which is also an eternal topic of landscape design. With the wisdom and subjective initiative of human beings, in the constant reflection and progress, we will continue to pursue the landscape design of "harmony between heaven and man" [2], which will definitely bring about tremendous improvement in human production, living space, and quality of life.

The ecological landscape of Cai fan and Longtanzhai in Taihu County studied in this project is not only an important expression of the settlement form, but also one of the research contents of landscape morphology. The main direction of this project research will be to study the content of the settlement ecological landscape from the multidisciplinary perspectives of geography, architecture and history (geography), anthropology, and art.

Methodologically,by collecting and sorting out the ecological concepts, planning layout, human space and other related materials embodied in the site selection process of the ancient village, it provides the necessary theoretical basis for the research of this topic. On this basis, we summarize the existing literature and historical data, and combine the ecological landscape of the ancient village with the design art, so that the ecological landscape of the ancient village has two characteristics – Artistry and Original ecology. The ecological concept, combined with the specific case of the ancient village, sums up the ecology of the ancient village, which is worthy of the place of contemporary landscape design learning. We have to go deep into the ancient villages of Cai fan and Longtanzhai in Taihu County to conduct on-the-spot investigations, understand the surrounding environment of the ancient villages, and the current development situation, take photos, visit the villagers' opinions and suggestions, and truly grasp the local accurate information, which complements the theoretical deficiency, in order to analyze the construction of ecological landscapes of regional ancient villages in a deeper level.

The richer the traditional resources of Cai fan and Longtanzhai ancient villages in Taihu County, the higher the historical, cultural, scientific, artistic, social and economic values, and the use of art as a medium to optimize and protect the ecological landscape of ancient villages, planning for ancient villages. In the process, it highlights its regional cultural connotation and continues the study of the ancient village culture. It analyzes the composition and characteristics of the ecological landscape of ancient villages, summarizes the paradigm of art to construct the ecological landscape of ancient villages, and explores the integration of native regional culture into landscape design to form a history with local style and characteristics.

## 2. Overview of ancient villages

### 2.1 Cai fan ancient residence

Cai fan is located in an ancient village in Tangquan Town, Taihu County, Anqing city, Anhui province. There are more than 300 well-preserved ancient dwelling houses. The building covers an area of more than 14,600 square meters and still houses more than 70 families. Since ancient times, Cai fan has enjoyed the reputation of "the scent of the book, the hometown of elegance."

Located in the deep mountainous area, the ancient village of Cai fan in the Jinying Village of Tangquan Town was built in the Ming and Qing dynasties. The buildings here belong to the typical style of the southern Fujian Huizhou and the hustle and bustle of the dragonfly (**Figure 1**). The antique is the building of the Ming and Qing dynasties in the southwestern part of the country. The typical example of concentration is also a model of the combination of the Chinese Huizhou architectural art and the Dabie Mountains. It is a "living stone" that reflects the twists and turns of the ancient villages in the Dabie Mountains (**Figure 2**).

The main building of Cai fan ancient residence on the three sides of the mountain is located in the northwest and south of the country. It is divided into ordinary dwellings, churches, and ancestral halls. The ordinary dwellings are two-storey brick walls maintained by the two-storey brick wall and still retain a large number of cultural connotations, engraving Yinglian plaque. The interior and exterior decorations are relatively simple, and the houses under the ridge are connected into a unified whole. The households are connected by roads, and a small river with bluestones surrounds the houses. The center of the church and the ancestral hall are

**Figure 1.**
*Blue tile roof of Longtan ancient village.*

**Figure 2.**
*Location of Longtan ancient village.*

made up of small bricks and small tile walls. The appearance is simple and elegant. The interior structure of the house is that the pillars cross over *fang* (Fang is a horizontal or vertical interpenetrating member that connects and stabilizes columns), and the pillars, beams and piggyback girders are hard and anticorrosive Castanopsis trees of local natural growth materials (**Figure 3**). These Huizhou art buildings such as the center of house and ancestral hall together with ancient ordinary dwellings constitute the Cai fan ancient dwelling complex. The building scale and volume are

**Figure 3.**
*Hu's ancestral hall.*

| Representative building | Construction time | Space layout | Remarks |
|---|---|---|---|
| 1. Xiatang hall | Kangxi 28 years (1689), during the Qing dynasty, | Setup courtyard, 10 wing room, stone door | Well preserved |
| 2. Zhongtang Hall | Yong Zheng, during the Qing dynasty, | Set up courtyard, wing room and wooden circular linen | Better preservation |
| 3. Shangtang Hall | 56 years of Qianlong (1791), during the Qing dynasty, | The main hall is supported by eight wooden pillars, with courtyard, and wing room | Well preserved |
| 4. The public housing of Chengxi | Kangxi 18 years (1679), during the Qing dynasty, | Liang Zhu structure, wood carving diagonal braces | Better preservation |
| 5. Weijia public housing | Yong Zheng 2 years (1724), during the Qing dynasty, | The structure of the house is simple, and the walls are covered with volcanoes | Wei Jia Gong is the common ancestor of the lobby; so, the whole residential building is centered on this house. |
| 6. Yuchun public housing | Qianlong period, during the Qing dynasty | The building is simple and elegant, wooden gate | The third generation grandchildren of Wei Jia; the buildings are well preserved. |
| 7. Former residence of Yin Yichen | Guangxu 7 years, during the Qing dynasty | The building is a typical Huizhou architecture, patio, wing room, horse head wall, and with an eight-character gate | Yin Yichen, who was in the 28th year of Guangxu, was appointed as the military cabinet in the 30th year of Guangxu. |
| 8. Yin's Ancestral hall | Qing Jiaqing 7 years (1802) | The building was made up of blue brick and blue tiles, carved corners, flats, inscriptions on the pilasters, and couplets. | Well preserved |

**Table 1.**
*Relevant data of representative ancient buildings in Caifan.*

huge. There are 650 houses with a building area of 13,800 square meters, which has a very high cultural value and the value of historical research (**Table 1**).

## 2.2 Longtan ancient village

### *2.2.1 Overview of Longtan village*

Longtan ancient village is located in Tangquan Town, Taihu County, deep in the hinterland of Dabie Mountain in the southwestern Yunnan, bordering Qianshan County in the east, connected to Siqian Town in the south, connecting Zhuwan Village in the west and Yuexi in the north. Longtan Village has a long history and profound cultural heritage. The village form, pattern, and concentrated distribution of the Qing dynasty buildings have extremely high historical value, cultural relic value, and artistic value. There are ancient populations of Hu's Ancestral Hall and Longtan Village with a history of more than 600 years. The Longtanzhai Nature Mountain is rich in water resources, such as small bridges, flowing water, and ancient trees (**Figure 4**). The environment is excellent, and the Huizhou architecture is profound, reflecting the model of harmonious coexistence between man and nature, and has a high academic value of cultural geography.

The Longtan ancient village buildings are located between the two mountains. The terrain is high and level, the mountains are beautiful, the streams are clear, and the air is fresh. The buildings in the ancient village are ancient and well preserved, which fully reflect the Huizhou style of the buildings and the southwest residential features and superb architectural craftsmanship. Since the establishment of Longtan ancient village has respected Confucian scholarship in education (paying special attention to Cheng Zhu's Neo Confucianism), the literary atmosphere is prosper, attaching importance to calligraphy and folk popular tune. Longtanzhai embodies its unique Huizhou cultural features from the geographical location and cultural phenomena of ancient villages, such as Huizhou merchant culture, social patriarchal culture, harmonious natural geomantic culture, and Confucian ethical culture of Cheng and Zhu Neo-Confucianism (**Figure 5**).

### *2.2.2 The cultural space of Longtan ancient village*

The "cultural space" belongs to the category of anthropology, and its concepts are numerous. The details are not the scope of this article. Mr. Liu Kuili: "according to the regional cultural characteristics of the same nature, the selection of traditional

**Figure 4.**
*The overall appearance of Longtan ancient village.*

culture is relatively complete and has a certain representativeness in terms of production methods, lifestyles, and conceptual forms. It has values, folk beliefs, and many specific cultural expressions" [3]. The scope of cultural space is as small as that of village society and as large as that of regional society. Its existence, in addition to the geographical form of settlement, focuses more on the concept of space in the cultural sense, that is, it can reflect the cultural expressions that are transmitted from generation to generation by a certain community and closely related to their lives. It is also the identification bond and cognitive space formed by the community in the process of historical evolution [4]. Therefore, the cultural space can be summarized in terms of common beliefs and cultural identities. Longtan ancient village has two clans', Hu clavus and Sri Lanka, long-term living in its habitat for a long time to form a cultural identity, and the patriarchal faith naturally is transmitted from generation to generation. This cultural identity and patriarchal religion are reflected in the concept of geomancy, landscapes, natural awe, and cultural awareness[5].

Longtanzhai pursues a Feng shui culture that is in harmony with nature in the architectural environment space and, at the same time, also pursues a landscape space. Longtanzhai always attaches importance to the feudal ethical culture of Cheng Zhu's neo-Confucianism, the culture of Huizhou merchants, and the Feng shui culture of landscapes. People in longtan village attach importance to commerce and respect culture. They know how to create elegant and peaceful living environment (**Figure 6**). Hillside structures built up and down the three-dimensional elevation of residential

**Figure 5.**
*Small view of Longtan ancient village.*

**Figure 6.**
*Part of Longtan ancient village.*

communities are with the formation of terraced mountain terraces and are combined with the ancient village of small ash shingles (for the traditional Chinese residential buildings above, which reflects a simple wooden components, walls, and other architectural elements and natural landscape). In addition, the mist in the mountains and smoke from kitchen chimneys made a natural landscape painting [5].

Herbert Reid once commented on the art: "art is always a symbolic dialogue. Once there is no symbol, no dialogue and no art" [6]. As an art form, the ancient village architecture, like other art forms, integrates symbolic and musicality, and contains the cultural spirit of a particular era. It shows the harmonious beauty in both spatial layout and geographical location. Being symbolic in Chinese traditional culture is an inertial psychological language that humanizes natural objects and formalizes them into ethical colors. Chinese architectural culture also has a unique symbolic cultural space, which is expressed in the singular number of the building's Taichung and Kaiji. This is verified by the celebrity ancient houses and the Huawu residential houses in the front of Longtanzhai's human landscape analysis. The symbolic nature makes the traditional architecture have a unique spiritual and cultural space. In the advanced stage of history, the ancient villages and houses have a strong monumental

| **Cultural landscapes** | | **Establishment of the era** | **Area** | **Social significance** | **Structural characteristics** |
|---|---|---|---|---|---|
| Hu ancestral hall | | Qing Jiaqing 6 years | 720 square meters | Religion | Civil structure, brick, stone, and wood carving decoration |
| Hui buildings | Shrine ancient houses | Qing Jiaqing years | 1600 square meters | Live | Stone for living, brick, adobe brick, wood, small gray tiles, and other material structures |
| | Bridge house ancient houses | Late Qing dynasty | 400 square meters | Live | Main structure of wood, adobe brick wall, and small gray tile roof |
| | Flower house ancient houses | Daoguang 8 years | 1782 square meters | Live | Courtyard of the falling wood structure |
| | Ancient house ancient houses | Qing dynasty Qianlong years | More than 2475 square meters | Live | Civil structure |
| | Ancient house on the house | Qing dynasty | More than 900 | Live | Civil structure and interior decorated with paint |
| | Hu million former residence | Late Qing dynasty | 1186 square meters | Commercial houses | Small brick and green brick structures |
| Former celebrities | | Guangxu early of Qing dynasty | 500 square meters | University mansion | The house is built in the wood and stone structure called *Si Shui Gui Tang* (with rainwater running from the roof to the courtyard) |
| Five blessing bridge | | Mid-Qing dynasty | 30 square meters | Ancient village only traffic arteries | Stone structure |

*Hu Xiaobing. Geographic Landscape Space and Tourism Planning of Longtan Ancient Village. Boletín Técnico, Vol.55, Issue 15, 2017, pp. 628-629.*

**Table 2.**
*Relevant data of representative ancient buildings in Longtanzhai.*

significance. The monumental value of Longtan ancient village is to store memory and structure history, and to transmit information to future generations through the construction of natural images to realize the communication between the living and the past (**Table 2**).

## 3. The ways for art to intervene in the planning of ancient villages

### 3.1 Constructing the ancient village native museum

At the ninth meeting of the International Museum Association in Paris in 1971, the French Georges Henri Riviere and Hugues de Varine made their debut in "Museum" in their speech. The word "eco" was prefixed with the word, and the concept of the eco-museum was born.

According to the definition of the Riviera, the eco-museum "is a tool that is nurtured, shaped, and operated by the public sector (local government) and local residents. Local governments provide librarians, equipment and resources, while local residents show their ambitions, knowledge and personalities. Power, so the Eco-Museum is a mirror of the local people to look after themselves, to find the image of the self, to seek an interpretation of the natural or human heritage in the field of life, and the life of the ancestors living in this field; Let visitors use a mirror to understand the local industry, customs and characteristics" [7].

The "ecology" of the eco-museum emphasizes not only the natural ecological environment but also the human ecological environment. Therefore, its philosophy is: "to protect and preserve the natural and cultural heritage in the original environment of the communities to which it belongs. In the Eco-Museum Different from the static collections in traditional museums, cultural heritage is a combination of dynamic and static display. They not only have meaning to researchers and tourists, but also have specific value, which is embodied in the cultural heritage of the museum. Natural landscapes, buildings, movable objects and traditional customs are the historical heritages of the original ecological nature in the long-term protection and continuous participation of the residents of their communities, and vividly reflect the social behavior of the community residents. Therefore, the ecology the construction of the museum should be based on the personal participation and personal management of the residents in the community, and seek the specific guidance of experts and scholars and the care and support of the local government. Only in such a situation can it be implemented to ensure its special vitality" [8]. Therefore, the regional nature of life, the national cultural heritage, and both the associated residents constitute the three elements of the eco-museum, while the originality, integrity, and authenticity are the three important principles of the eco-museum.

### 3.2 Construction of cultural ecotourism

Cultural ecology is the premise of the protection of ancient villages. The uniqueness of the ecological landscape of ancient villages is, in the final analysis, the uniqueness of cultural ecology. Every ancient village grows and evolves under specific ecological conditions. At present, the historic ancient village architecture is an important human cultural heritage, and also an important part of tourism resources. The construction of eco-tourism has positive significance for the ancient village to present its glorious side to the world. First, eco-tourism development has created a new living space for ancient villages, which makes it reflect the use value and is conducive to its protection. Second, the development of eco-tourism has expanded the audience of ancient villages and cultivated new ones. The audience group is conducive to improving the popularity and influence of the ancient villages. Third, the development of eco-tourism can bring certain economic benefits to the ancient villages, help them to establish their own maintenance functions, and achieve a virtuous cycle of "protection-development-protection". Fourth, ecotourism development has prompted government departments to increase the efforts to rescue and protect the ancient cultural heritage of the human village [9].

The ecological environment of Taihu County is superior. There are many famous scenic spots in and around the county, such as Tianzhu Mountain, Sanzu Temple, Huating Lake, Xifeng Temple, etc., as well as the ancient villages of Cai fan and Longtanzhai constitute a multipoint line tour (**Figure 7**). In terms of geographical location, Cai fan and Longtan Village are located on the routes of these scenic spots. On the basis of maintaining the original ecological environment, we should build tourism-related projects, such as homestays, ecological farms, cultural and creative centers, etc., so as to integrate accommodation, tourism, learning, food, and leisure in one. Residence accommodation is a mainstream trend of tourism development. It is based on the concept of green ecology, using its own built environment and local ecological environment, natural landscape, humanities, and special resources to provide visitors with a cultural accommodation

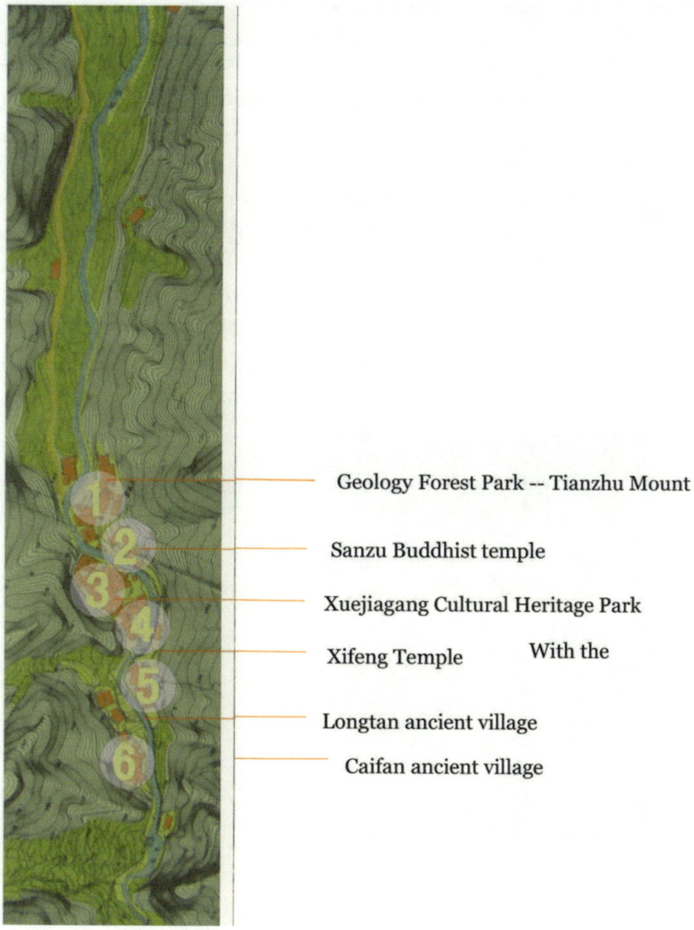


**Figure 7.**
*Tourist route.*

experience [10]. In terms of interior decoration, the ancient villages in Taihu Lake can be decorated with decorative paintings, such as the history of ancient villages, the customs, the characteristics of scenic spots, local specialties, field minors, folklore traditional dance, self-entertainment dance, sacrificial dance, etc., to adorn the inner wall for tourists' appreciation. In this way, not only the ecological landscape of the ancient villages has been improved, but also the tourists have a comprehensive understanding of the cognition and protection of the ancient villages, which is a kind of living protection.

### 3.3 Art mining to build the cultural consciousness of ancient villages

With the rapid advancement of globalization and urbanization, Chinese towns and villages are changing dramatically [11], and the identity dilemma of cultural identity is also approaching us. On the one hand, we need to quickly integrate into the modern civilization system of the world by means of modernization. On the other hand, we must establish our own cultural identity and maintain our own cultural identity.

Some current cases of art resident projects in some ancient villages in the country are worthy of reference for ancient dwellings in the Anqing area. For example, Xiaozhou Art Village of Guangzhou is a familiar case (**Figure 8**). It started from the artist renting idle houses of the village as a studio. The tremendous changes taking place in the ancient villages of the water town fully prove that the power of art can inject new vitality into the ancient villages and achieve sustainable development without over-commercialized tourism.

When a large number of artists came to the same place to make artistic creation, the villages with excellent mountains and waters gradually became the base of art relocation, which is also a way to activate the ancient villages. In the Cao fan and Longtanzhai villages of Taihu County, there are many ancient buildings and cultural relics in the Qing dynasty. If the two villages are based in Anhui and surrounding provinces and cities, then they are oriented to the whole country and take artistic forms such as sketching, artistic creation, photography, etc., carry out cooperation with famous colleges and universities, and promote various forms of art exchange activities such as art sketching, pen meeting, exhibition and so on,

**Figure 8.**
*Xiaozhou art village.*

and strive to build a bridge for artists and painters. The artistic creation base of spokespersons and brokers. Initially, the villagers may have been indifferent to the establishment of sketch bases in China's art colleges and universities. However, when they saw the employment opportunities and dissemination effects of the scheme, they actively devoted themselves to the relevant work and passed on this traditional folk custom.

## 4. Methodology

This topic uses art as a medium means in the paradigm study of the ecological landscape of ancient villages. In the practice of building the ecological landscape design of the village, its purpose is to strengthen its regional cultural connotation and continue the study of the ancient villages. The research methods such as literature collection, field research, and inductive summarization are used to explore the carriers, techniques, and strategies of art in the construction of ecological landscape in ancient villages. The topic is to take Cai fan and Longtan Village as the research object, then building the theoretical system foundation. At the same time, the topic analyzes the ecological landscape of the two ancient villages, in order to further explore the practical application of the expression of art media in the ancient village landscape, so as to reach the paradigm theoretical framework for the construction of the ecological landscape of ancient villages.

From the above analysis, we will create Cai fan ancient village and Longtanzhai as an attractive and natural ecological space that promote the two ancient villages' sustainable development. The framework is shown in **Figure 9**.

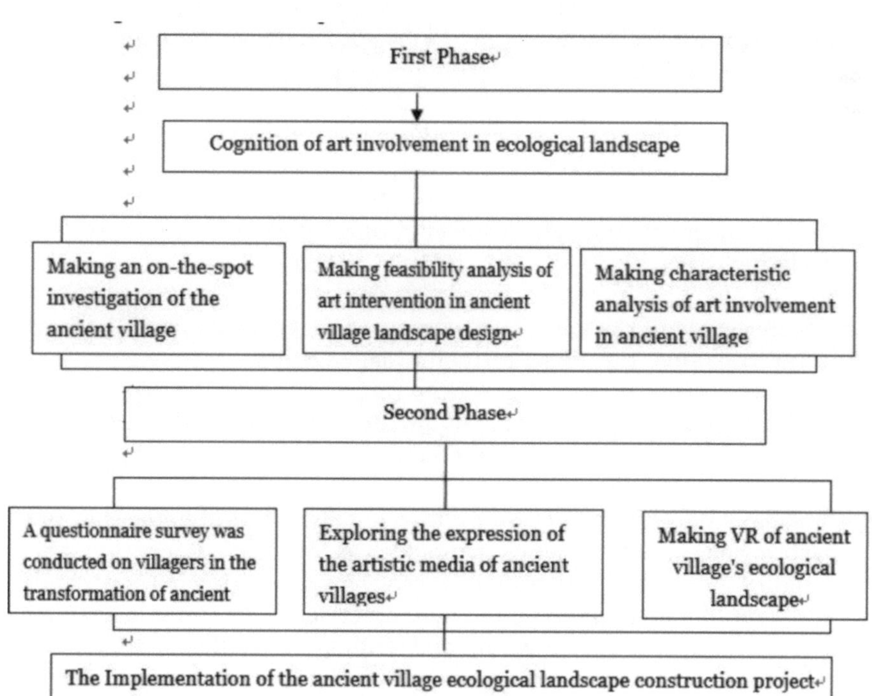


**Figure 9.**
*Research method framework.*

## 5. Art involved in the construction concept of Caijing and Longtanzhai ancient villages

### 5.1 Pay attention to the geographical environment and ecology of ancient villages

Due to the mountainous conditions, the Cai fan and Longtanzhai ancient villages in Taihu County will be located in the undulating mountains, and the central location of the village will also be uneven due to the trend of the mountains. The villagers will build houses according to the trend of such ups and downs. Such terrain effects will occasionally play some unique forms of planning. There is not only a change in terrain difference between houses and houses, but also a rich and varied ecological natural landscape. The ancients built high walls behind the garden of the courtyard, cleverly using the height difference, so that the strong visual effect would make the bottom of the upper courtyard appear infinitely extended and become the center of vision [12].

The landscapes of the Cai fan and Longtanzhai ancient villages in Taihu County are backed by mountains. In fact, the location of such a house, such as facing the sun, relying on green water, leaning against the mountains to avoid the wind, is the best natural nutrition for life. "Laozi" calls "everything is negative and yang." These ancient villages can clearly reflect the rich experience of the ancient villagers and the awe of the ecological nature.

### 5.2 Consider the safety and commercial circulation of residents

The construction of the ecological landscape of ancient villages must consider the safety of living. This aspect must be considered very thoughtfully, and it is necessary to solve this problem in a comprehensive way. Because the security of residents in ancient villages is the foundation for the development of a village, and it is the top priority for the development of villages. So, the landscape construction meets the following points:

- First of all, according to the terrain of the ancient village, commercial circulation is the first consideration.

- Second, we must have a good natural environment and sufficient sunshine to form a living space.

- Finally, the layout of the village, the courtyard organization, and the mountain line echo each other, forming a better drainage system.

### 5.3 The ecological landscape planning of ancient villages should be based on the modern environmental planning and Feng shui design

Feng shui has a long history in China and gradually matured in the Han and Tang dynasties. Feng shui includes elements such as form, law, stagnation, and fate. The most important of these is that there is a strong superstition in theory and fatalism, but the method has certain scientific and practical significance in analyzing topography, small physical environment, and guiding the modeling of architecture [13]. When Feng shui was in ancient China, it was widely spread among the people, and its influence on traditional Chinese dwellings was extremely obvious.

The application of Feng shui to the site selection of the village actually includes many specific aspects, such as climate, ecology, topography, geography, landform, landscape, etc., as well as the interaction between the factors, as well as taboos and shortcomings. In other words, Feng shui emphasizes the comprehensive evaluation of various factors. Feng shui theory has an important guiding role in the landscape of ancient villages. Environmental science is an important part of Feng shui theory, and it also has certain scientific ideas. The theory of Feng shui will also explain the part of the gods, which is also a reflection of the social environment at that time. It is stated that the ancients will fully consider the theory of Feng shui in the site selection of houses. It is more obvious in the wealthy people. In the construction of the building, the location and spatial relationship of the nearby mountains and rivers clearly illustrate this point.

### 5.4 The structural layout design paradigm of the ancient village should reflect the original ecology

The structural layout of the ancient villages of Taihu and Longtan Village is the wisdom that reflects the wisdom of ancient residents. If the housing of a household needs to take into account all aspects of Feng shui, environment, site selection, safety, etc., then the formation of a village requires the sum of the required factors for each household, and it is necessary to adjust each household, the needs of the household, and the process of integration. Therefore, the layout of the ancient villages in Taihu County is also the content of the landscape planning we have talked about in the contemporary era. We can learn more ecological and smarter ancient thoughts and cultures from the structural layout of ancient villages.

## 6. The social value of art intervention in the construction of Cai fan and Longtanzhai ancient dwellings

First, it will contribute to the inheritance and reproduction of the ancient village space in Anqing.

The ecological landscape of ancient villages constructs the core concept of its paradigm. First, the village is the carrier of traditional memory. On the one hand, the countryside is the carrier of traditional memory. As a concrete operator, the designer needs to follow the concept from beginning to end: how to participate in the existing courtyard space with the new lifestyle, how to integrate modern art into the past and present life, and how to carry the new art life with the historical space. On the other hand, when artists participate in rural construction, they must understand that local farmers are the main body, not the self. The comprehensive approach of "moving, dismantling, repairing, supplementing, and decorating" was adopted, thus leaving the era of traditional ancient villages such as geography, ecological landscape, and construction features.

The second is to contribute to the protection of the ecology and the construction of the landscape.

The ancient village in Anqing is located in the north of the Yangtze river. It is surrounded by mountains and rivers and has beautiful natural scenery. The ancient villages pay attention to their ecological protection during the landscape construction process. The optimization of the river network and the beautification of the landscape are well maintained. Doing appropriate optimization on the basis can be said to be everywhere.

The third is to help protect culture, combine local elements, and highlight the regional cultural characteristics.

We will pay attention to the protection of cultural relics and the repair of ancient buildings, and strictly protect the original habitat of ancient villages and tap the regional potential. The ancient village culture has the dual attributes of material and nonmaterial, as well as its interrelated, unique, and recognizable characteristics. It is rooted in all aspects of rural production and living. Whether it is traditional agricultural production methods, folk cultural activities, local architecture, agricultural production landscapes, etc., it is the expression and bearing form of regional cultural symbols. The production and lifestyle, folk customs, and regional environment in Anqing area are different. The characteristics of architectural form, shape, site selection, material selection, decoration, and crafts are also slightly different, forming a unique concept of local architecture. The structure and environment of the village structure with various local conditions and various forms. These unique cultural elements of ancient villages are the material and spiritual carriers that can best reflect the characteristics of local culture. They are the important core to maintain and reflect the local characteristics and style of the villages, and to arouse the people's attribution to the local, regional, and national culture being an important way of feeling, identity, and pride.

The fourth is to help create an ecological landscape that can live and swim.

The construction of ecological landscapes in ancient villages and the development of eco-tourism should be based on respecting the historical evolution of the countryside, folk customs, traditional customs, etc., integrating history and modernity, and integrating local culture and ideas into contemporary rural construction and transformation. People create a rural production and living space with a better local culture and a higher quality of life. Maintaining and inheriting local culture is not only conducive to strengthening the country's own characteristics and ecological construction, but also effectively promoting the sustainable development of rural tourism, tourism, and experience. The intervention of art, on the basis of maintaining the original ecology of the ancient villages, makes the ancient villages more accessible by the overall concept of repairing and reconstructing.

## 7. Conclusion

"Art participation in the construction of ancient villages" is not an end, but a working method. It is not a new thing to say. It has been successfully applied in many countries since World War II. This kind of work is a kind of cultivation, not a short-term project behavior, but a long-term, endogenous, and sustained vitality. All the art will become a document in the construction of ancient villages and become part of the collective memory of ancient dwellings.

The ecological landscape of Cai fan and Longtanzhai ancient villages in Taihu County is not only an important expression of the settlement form, but also one of the research contents of landscape morphology. The main direction of its construction will be to study the content of the settlement ecological landscape from the multidisciplinary perspectives of geography, architecture and history (geography), anthropology, and art.

Due to the acceleration of the modernization process, the destruction speed of some ancient villages is staggering, and the cultural heritage of the ancient villages left for the next generation is gradually less. Based on this, the human landscape and ecological landscape of the ancient villages are constructed by the art media to construct the landscape of the ancient villages. On this basis, the dual effects of the protection and transformation paradigm are achieved, and the ancient villages are enriched by the premise of maintaining the original ecological landscape. With the ecological connotation, the ancient villages have a liveable and accessible landscape. In our best practices for achieving sustainable development, we strive to balance economic, environmental, and social factors to ensure resource conservation and environmental protection for resident who can benefit from it; at the same time, the local government can also be inspired by the construction paradigm, which could provide constructive advice for them.

## Acknowledgements

This project is funded by humanities and social science research key project in Anhui province 'Research on the Development of Opera cultural and creative Products from the Perspective of Huangmei Opera' (SK2018A0344).

## Author details

Hu Xiaobing[1,2] and Zhang Yalin[1*]

1 Jingdezhen Ceramic Institute, Jingdezhen, Jiangxi, China

2 Anqing Normal University, Anqing, Anhui, China

*Address all correspondence to: 5884286@qq.com

## References

[1] Peilin L. Ancient Village: Harmonious People's Space. Shanghai: Shanghai Sanlian Bookstore; 1997. pp. 45-49

[2] Zhu X, Wang C, Wang J. Jin Merchants Residence. Beijing: China Building Industry Press; 2009. pp. 50-56

[3] Peilin L, Dong S. Spatial imagery of ancient chinese villages. Geographical Research. 1998;**17**(1):31-38

[4] Kuili L. Discussion on the issue of cultural and ecological protection zones. Journal of Zhejiang Normal University (Social Science Edition). 2007;**32**(3):9-12

[5] Hu X. Geographic landscape space and tourism planning of longtan ancient village. Boletín Técnico. 2017;**55**(15):629

[6] Abercrombie S, Yucheng W. Architecture as Art. 1st ed. Tianjin: Tianjin University Press; 2001. p. 129

[7] Freeman H. The museum without walls. International Museum. 2006;**58**(2):55-59

[8] Hu C. The practice of ecological museum theory in Guizhou. Chinese Museum. 2000;**63**(2):61-65

[9] Ma J. The concept and method of cultural and ecological protection. Fujian Art. 2011;**195**(5):29-33

[10] Gaelic Y. The Space of Communication and Communication. Beijing: China Construction Engineering Publishing; 2002. pp. 16-20

[11] Zhihua C. Chinese Native Architecture. Chongqing: Chongqing Publishing House; 2001. pp. 25-27

[12] Shen F, Jie L. The Ecological Outlook of Ancient Chinese Architectural Environment. Hubei: Education Press; 2002. pp. 19-26

[13] Hao C, Zengfeng X, Stupid Z. Hancheng Juejia Village Regional Fore Analysis. Anhui: Agricultural Science; 2011. pp. 27-30

Chapter 4

# The Impact of Landscape Reclamation on Mangrove Forest and Coastal Areas in the Niger Delta, Nigeria

*Aroloye O. Numbere*


## Abstract

Coastal area is in serious danger from land reclamation in the Niger Delta, Nigeria. This is because of land expansion activities such as urban development. Landscape reclamation is intended for urban city expansion, road construction, housing project, crude oil exploration and sand mining. Reclamation is carried out by both government and private developers. The government sometimes forcefully acquires coastal areas from the native community, remove the mangrove forest and sand fill the area in other to establish projects beneficial to the public. Private investors reclaim coastal areas to execute private business that would boost their economic fortunes. Oil companies clear coastal forest and set up oil wells and pipelines in swampy locations. Increasing population in small communities had also led to the reclamation of coastal areas to create room for the construction of houses to accommodate more people. However, many land reclamation activities are not development-centered, but business-centered. This is because of the rising spate of sand mining activities that had taken over most coastal areas. Sand mines are often abandoned after some years of operation. Reclamation is done without proper environmental impact assessment. This situation had led to the loss of many species.


**Keywords:** Niger Delta, urbanization, invasive species, hydrocarbon pollution, mangrove, exploration, seismic activities, oil spillages, pipelines

## 1. Introduction

Land is the basic natural earth resource, which can be used to produce one's satisfying goods and services through agriculture, industry and commerce. Land in the Niger Delta area is in large demand because of traditional and cultural beliefs attached to the possession of land [1]. Land is regarded in this area as the best form of asset investment because of the prestige attached to its ownership. The cost of land appreciates and has no depreciative value. But a land or coastal area that has been plundered and polluted may loss its economic value [2]. The aim of acquiring land varies, but is mainly for possession and or production of goods and services to satisfy socio-economic and socio-cultural needs. The following are natural resources found in land in Nigeria: columbite, dolomite, gold,



IntechOpen

tin, iron ore, limestone, silver, uranium and crude oil. The crude oil resource is mainly present in the Niger Delta region of Nigeria, which is the centripetal force on which the nation's economy revolves [3]. Land is highly priced and many have died for its sake through fratricidal wars and communal clashes because of its scarcity. This situation had precipitated the uncontrolled reclamation of coastal lands [4]. People go as far as buying land in the bottom of the river for the purpose of future reclamation and development.

Land policy in Nigeria is influenced by government and culture. This is encapsulated in the Land Use Act promulgated in 1978, which states that all lands in Nigeria belong to the Federal Government, this include but not limited to underwater land that is 200 nautical miles from the shore. The Governors of each states of the federation hold the land in trust for the Federal Government, and are entitled to be the sole signatory of all certificate of occupancy (C of O) before a piece of land is legally owned by an individual or group of individuals. Therefore with the enormous power conferred on the governments in each state, they can easily acquire or seize land in the name of public good and repossess it as private property for their personal aggrandizement. Before coastal lands are to be acquired proper environmental impact assessment (EIA) studies need to be conducted to develop proper management plan aimed at forestalling loss and extinction of common and rare species. But in most cases no proper EIA or landscape assessment is carried out [5]. It is the responsibility of the Government to protect the land against illegal landscape reclamations. However, in many communities coastal lands are protected traditionally if they have cultural significance to the people [1]. Despite the negative consequences of coastal land reclamation to aquatic species, it can add some economic value if used for non-intrusive agricultural activities. Nevertheless, land utilization is influenced by nature and characteristics of soil, soil moisture and temperature, topography and land location, flora, fauna and climate. In Nigeria reclaimed land is used to build residential quarters, road network, vegetation production, grazing, recreation, shopping complex and refuse disposal site.

Land reclamation is also known as land fill, it is the process of creating new land from ocean, riverbeds or lake beds. It is the returning of lands to an improved state. It is also referred to as the process of improving lands to make them suitable for more intensive use. Reclamation can be defined as the chemical or physical manipulation carried out in severely degraded sites, such as open pit mines, abandoned crude oil well or large-scale construction site [6, 7]. Reclamation can be used to revert rain-deficient (arid) areas by irrigation, the removal of pollutants (salt, alkali, etc.) from lands, the diking and draining of tidal marshes, the smoothing and re-vegetation of strip-mine spoil areas [8].

Historically, reclamation meant irrigation projects that brought wetlands and deserts (considered useless wastelands) into agricultural production. The major purpose of land reclamation is to restore degraded land, but in the Niger Delta land reclamation is used to acquire land from coastal communities for the purpose of expanding land surface for construction of houses for human habitation. Land reclamation in the Niger Delta passes through five phases, which in all ramifications affect the environment. These phases include: (i) deforestation of mangrove forest (ii) consolidation of swamp with bulldozers, (iii) pumping of white sand from the sea bottom unto shore, (iv) sand filling of reclaimed land, and (v) construction activity e.g. roads, buildings, industries and parks.

Land can either be physical or economic. Physical land covers all the earth including land surface, sub-surface, under water and super surface (atmosphere). Economic land on the other hand, is part of physical land that can be used to produce economic commodities for man's satisfaction [2]. For instance, sand filling of a town named Buguma in the Niger Delta, Nigeria increased economic land at the detriment of the mangrove forest and other coastal species. Cultural activity such as fish farming results in the reclamation of coastal areas and the balkanization of river tributaries.

## 2. Reasons for land reclamation in the Niger Delta

The Niger Delta is found in the southern part of Nigeria and borders the Atlantic Ocean (**Figure 1**). Mangrove vegetation is the dominant species found at the interface between the land and the sea. Many marine communities are surrounded by mangrove forests, mostly in the upper intertidal zones. Population increase has led to the migration of people towards the coastal locations. The causes of land reclamation in the Niger Delta are grouped into two: (1) direct and (2) indirect causes. The direct causes include: (i) land expansion, (ii) construction activities, (iii) land acquisition and (iv) succession (land-forming activities of mangroves). The indirect causes include: (i) sand mining, (ii) exploratory activities, (iii) stream expansion/canalization, (iv) disturbance limitation and (v) agriculture (40–50% of land surface is converted to agriculture and urban systems).

### 2.1 Direct causes

#### *2.1.1 Land expansion*

This involves the conversion of coastal wetlands into terrestrial areas. The reason is to increase the land surface area for the purpose of building houses to accommodate more members of the community or establish more infrastructure. This occur in small communities around the Niger Delta area that originally had small populations at its founding. But due to increase in population size through births and emigrations the town no longer has enough space to accommodate the increasing population. This situation thus necessitated the reclamation of coastal areas to create more space for human habitation. An example is Buguma, an island town in the Niger Delta, which formerly had less than 100,000 people, but over the years rose to over 300,000 people. This situation made the local authority to cut down

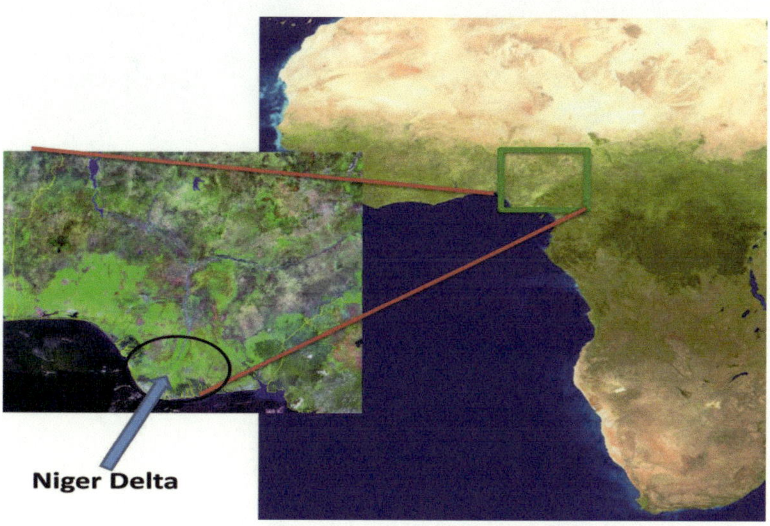


**Figure 1.**
*Map of Niger Delta Nigeria (circled) bordering the Atlantic Ocean has the biggest mangrove forests in Africa.*

and sand fill a mangrove forest measuring about 4.2 million $m^2$ in 1984 (**Figure 2**). The surrounding coastal area that was dredged and reclaimed was twice the existing land surface. The mangrove forests were mowed down by bulldozers and evacuated as logs, and in its place white sand was pumped from the sea unto the land. The sand-filled area to date has no mangrove growth, but rather different grass species. Presently buildings have been erected on the site, which are occupied by some people. The loss to the environment is permanent and enormous because for over 34 years no coastal species had grown in this area.

The second example is another site known as Borikiri in the outskirts of Port Harcourt, the capital city of Rivers State. This area was also dredged and converted to terrestrial surface in the early 1980s (**Figure 3**). So far in this location thousands of houses had been built, which houses over half a million persons. The implication of these landscape reclamations is that the destruction of mangrove forests lead to a colossal loss of ecosystem services to the environment. Anthropogenic activities around the coast is detrimental to its sustainability because of the addition of pollutants and contaminants [5].

### *2.1.2 Construction activities*

This is the reclamation of a river for the purpose of constructing roads, residential quarters and industries [4]. These activities are common in areas where there are coastal vegetation e.g. mangrove wetlands. The swamps are scooped away or reinforced with sand and concrete in other to produce a hard surface on which to construct foundations of buildings. Swampy soils are also locally reinforced with hard soils known as "chikoko" and left for some years to solidify before houses are

**Figure 2.**
*Dredged and sand filled areas in Buguma, Niger Delta, Nigeria. The white patch indicates the sand filled area while the green patches indicates mangrove forest that is still standing. The white patches sum up to give a total of 4.2 million $m^2$ of mangroves removed in 1984. This estimate was made using Arc GIS [9].*

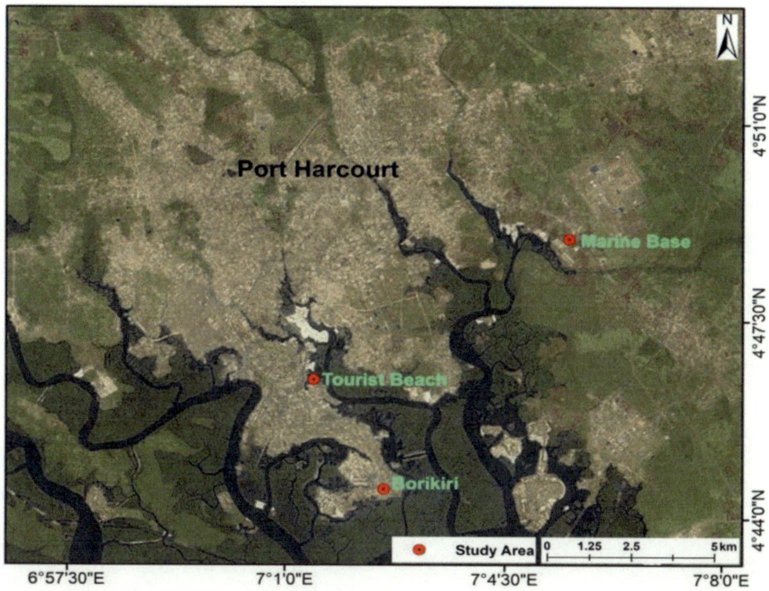


**Figure 3.**
*Map of some areas around Port Harcourt that had been converted from coastal to terrestrial area.*

built on it. Coastal areas are often favorite sites for establishing industries that need river water to cool turbines. Industries also flock around coastal areas because of the ease of transportation of goods and equipments.

Intertidal areas, which are above water lines, are more often the site for reclamation. These areas are naturally covered with water during high tides, but are filled with soil and consolidated to prevent tidal flooding during high tide. The solidified area is used as a platform for construction activities. Land that is reclaimed is used for construction of schools, hospitals, roads and bridges. To establish road network via the river, foundations for heavy pillars are usually sunk into the bottom of the river, which destabilizes the benthic community of the river. Areas close to the shore are also sand filled to connect bridges. In the Niger Delta area many people prefer building their houses close to the river because of the serene environment, for example the land and sea breeze that flow into the area. Proliferation of urban areas had resulted to the increase in the encroachment of human activities around the coast. Over population is a major factor that had led to the extension of cities beyond coastal limits. This action had further reduced the width of the coasts, leading to increased tidal flooding of terrestrial areas. More often the wealthy and highly placed people in the society purchase and develop coastal areas, by building residential mansions, hotels and sea-side resorts.

### *2.1.3 Land acquisition*

It is the forceful or legal takeover of coastal area by individuals or government officials. Highly placed individuals prefer constructing their houses along coastal areas. Through the use of police power they acquire and take over mangrove forest they bulldoze the forest, dredge and sand fill the site, which they allow to lie fallow for some years or reinforce and develop immediately. Coastal sites are preferred by land speculators because they usually off-city limits and isolated from the rest of the population.

#### *2.1.4 Succession as a primer for land reclamation*

Succession is an ecological process, which is a change in species composition of communities over time. It is the result of abiotic (physical and chemical) and biotic agents of change. Mangroves are regarded as land forming organisms [10]. Thus mangroves are natural land reclaimers. This is because their adventitious root system traps sediments during tidal flow, and accumulate it over a long period of time. Presence of sediments leads to the gradual formation of terrestrial areas, which becomes the habitat for plants. Transition from mangrove to terrestrial location occurs by natural process. Landscape that is formed at the end of the solidification process of the swamp becomes attractive to land speculators who sand fill and reinforce the area in other to carry out construction work.

### 2.2 Indirect cause

#### *2.2.1 Sand filling*

Almost every coastal area in the Niger Delta has had one abandoned project or the other, one of which is sand dump. Sand mining activity is a lucrative business in the Niger Delta and is embarked upon by both private individuals and government officials. Local sand mining is done manually by the digging of sand from the bottom of the river during low tide and conveyed ashore in hand dug canoes. Sand mining is also done with more sophisticated machines, where pumps are used to convey sand from the bottom of the river unto land via long pipes. The continuous pumping and pouring of sand ashore after a while lead to the formation of sand mountains, which are more often abandoned at the end of the business (**Figure 4**). Deposited sand is usually evacuated by trucks to buyers. The environmental problem of this practice is that the sand dumped by the shore smothers plants and animals around the area and changes the coast from a marshy to a sandy area.

#### *2.2.2 Exploratory activities*

Oil and gas exploration is the main exploratory activity that occurs in the Niger Delta region. It occurs at on-shore or off-shore locations [3]. Before exploration

**Figure 4.**
*Former mangrove forest that was converted to sand fill in Buguma, Niger Delta, Nigeria.*

the site is prepared through pilling, dredging, sand filling, grading and concreting. Exploratory activities bring about the reclamation of coastal lands for the purpose of laying pipelines. It also involves the establishment of well heads and booth camps. During exploration the coast is first dredged, cleared of vegetation and reclaimed in other to create a plat form for mounting pillars that carry oil pipelines and well heads. This occurs during off-shore drilling in mangrove swamps. Surrounding areas of the swamps are usually dredged and sand-filled to enable it to support the mounting of crude oil well head and the laying of pipelines (**Figure 5**). Concrete bases are constructed on this platform to carry the pipes from the drilling point, which may be at an off-shore or on-shore site to the refinery. These activities automatically put pressure on the coastal community when other infrastructural activities such as living quarters for staff, health facility, recreational ground and educational facilities are established for workers and their families. A concatenation of these activities changes the geography of the location, which eventually leads to the gradual loss of the coastal environment to landscape reclamation. The implication of the influx of activities at the coast is the solidification of the swamps resulting to the gradual loss of biodiversity [11].

### *2.2.3 Stream expansion and canalization*

This is a situation where rivers near ports are expanded to accommodate large ocean going ships. Canalization is done to enable ships to berth at the wharf, without running aground. During canalization, earth moving dredgers are deployed to excavate the benthic soil. The scouring of the river bottom crushes and destroys a lot of organisms such as shell and non-shell fish community in that location. The dredged soil that is evacuated from the river bottom is dumped on land surface and smothers coastal vegetation such as mangroves. Accumulation of dredged spoil converts the swamp to terrestrial area and also changes the physico-chemistry of the coasts [3]. Expansion of creeks allow for navigational activities, commerce and construction. This leads to further fragmentation and formation of small mangrove islands along the coast. Changes in land form lead to changes in the biogeography of the area. Terrestrial land form eventually evolves from an aquatic environment after solidification.

**Figure 5.**
*Crude oil pipeline and well head situated in reclaimed coastal area in the Niger Delta, Nigeria.*

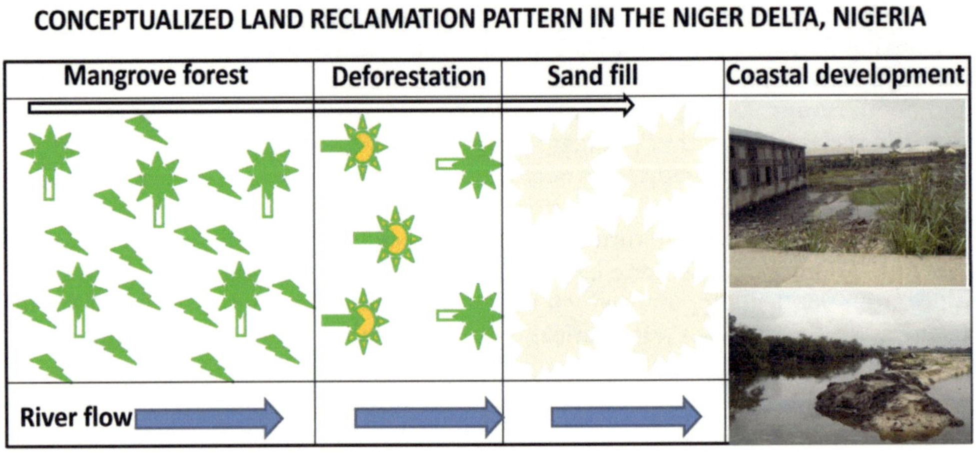


**Figure 6.**
*A conceptualized reclamation pattern in coastal communities in the Niger Delta, Nigeria.*

### *2.2.4 Disturbance*

Invasion of foreign species in coastal areas is another precursor to land reclamation activity aimed at remediating the sites in the Niger Delta region. Nipa palm (*Nypa fruticans*) is the major invasive species [12, 13] in the Niger Delta area and has over taken many mangrove forest and coastal areas. The problem is that mangroves provide numerous ecosystem services to the local people while the palms provide no ecosystem service to the people. Because of the non-use of the palms they had been removed from several locations with the aid of swamp buggies as a means of mitigating their negative environmental effect. The entry of big machinery into the swamp had added more stress to the coast by destroying many soil-dwelling and benthic organisms. The problem of this action is that the nipa palm forest after being cleared is repossessed by the government for infrastructural developments, which rather than reduce the negative effect worsen the situation of the coast. An example of this pattern of landscape reclamation is found in a place called Eagle Island, Niger Delta where the mangrove forests were cut down and sand filled and later the area was used for developmental project (**Figure 6**).

### *2.2.5 Agriculture*

Agricultural activities such as rice paddies and aquaculture are embarked upon in swampy coastal locations [14]. These activities are economically beneficial to farmers, but inimical to the environment. This is because such activities lead to the clearing of the mangrove forest and other coastal vegetation. It also changes the coastal structure and lead to flooding of upland areas. The soil chemistry is also affected by the manures and fertilizers that come from the farm and the waste from fish ponds. A combination of these activities leads to the acidification of the river causing fish deaths and physiological and reproductive problems of numerous marine organisms.

## 3. Impact of land reclamation on mangrove and coastal environment

Land reclamation impact the coastal environment in several ways (**Table 1**).

| Potential impact | Site preparation | | Dredging | Sand mining | Sand filling | Construction |
|---|---|---|---|---|---|---|
| | Deforestation | Clearing | | | | |
| Air emissions | ☆ | ☆ | ☆ | ☆ | ☆ | ☆ |
| Noise & vibration | ☆ | ☆ | ☆ | ☆ | ☆ | ☆ |
| Impacts on mangroves | ☆ | ☆ | ☆ | ☆ | ☆ | ☆ |
| Reclamation | ☆ | ☆ | ☆ | ☆ | ☆ | ☆ |
| Impacts on wildlife & coastal life | ☆ | ☆ | ☆ | ☆ | ☆ | ☆ |
| Geology & hydrogeology | ☆ | ☆ | ☆ | ☆ | ☆ | ☆ |
| Impacts on soil & water | ☆ | ☆ | ☆ | ☆ | ☆ | ☆ |
| Health impacts | ☆ | ☆ | ☆ | ☆ | | |

**Table 1.**
*Potential environmental impact of land reclamation in the Niger Delta, Nigeria.*

### 3.1 Environment

#### *3.1.1 Site vegetation clearing*

Before reclamation the site is cleared, which leads to the destruction of native plant species within the reclaimed area [3]. This could lead to permanent loss of mangroves or other plant species of economic/medicinal importance and habitat for marine organisms and wildlife and their emigration to unaffected areas, thereby upsetting the ecological balance. During clearing heavy machinery fell trees. This further denigrates the soil structure and converts the aquatic environment to marshy environment. The swampy environment is a mixture of mud and water, after a long period of perturbation the swampy soil gets solidified and changes to a terrestrial environment.

#### *3.1.2 Increased erosion of the cleared areas/river banks*

Once the bulldozer rolls in to mow down the mangroves, it creates depressions for tidal pressure to wash in ashore. This is because the adventitious roots of the mangrove forest serve as tidal breaks. The presence of mangroves along the coast stabilizes and reinforces the soil against erosion. Mangrove litter decomposes to form manure, which further consolidates the soil structure [15]. Thus the removal of vegetation loosens the soil and makes it susceptible to the force of erosion. In the same vein the wheel of the bulldozers fragments the soil particles and makes it porous and prone to leaching or wind erosion.

#### *3.1.3 Increase access for hunting and logging*

Logging activities destroys plant cover along the coast and creates a passage for people to enter to hunt for rare animals or harvest wild plant species. Similarly, reckless and indiscriminate hunting in opened forest reduces population and thus reduces species diversity, which may lead to extinction of some species.

#### *3.1.4 Changes in topography of sand filled area/river bed and dredged areas*

This could lead to the death of soil dwelling organisms. Changes in river channel via canalization or dredging changes the river system by increasing the length and breadth of the river which affects the flow dynamics and hydrology of the river [3]. This situation affects the fish survival and population. Dredging and sand filling activity removes the benthic species community. It also deepens the sea bed, which affects the water level. Decrease in water level affect intertidal level especially when areas along the shore receive no water supply and become dry. This leads to the death or migration of amphibious species along the coast. The dehydration of the intertidal areas cuts off dissolved oxygen and food supply to aquatic species.

#### *3.1.5 Increased turbidity*

This could lead to the reduction in species composition and diversity of aquatic resources. It could also lead to fish kills and smothering of aquatic organisms. Increased turbidity leads to the blockage of light beyond water surface, thus hampering photosynthetic activities below the water surface such as the benthic region.

#### *3.1.6 Disturbance of aquatic life*

This includes zooplankton, phytoplankton, benthic organisms and fisheries. It also affects soil-dwelling organisms from adverse impact on water and soil quality. The felling of trees and the uprooting of their stumps leads to soil fragmentation. Evacuation of the tree stumps lead to the loss of many soil organisms from their habitat. Sand filling of the coasts lead to the burial of millions of species. Loss of trees leads to increased soil erosion and increased sedimentation of river.

#### *3.1.7 Ground water/soil quality could be impaired by leachates from generated dredged soil*

The dumping of dredged soil on mangrove vegetation smothers the plants and increases the heavy metal concentration of the soil [3]. This affects the survival of soil-dwelling organisms by increasing the acidity and alkalinity of the soil. Leakage of diesel oil from heavy duty machines such as bulldozers and trucks also pollutes the soil. Similarly, the contamination of surface and ground water with used engine oil can pollute the ground water aquifer, thus affecting neighboring community that drink water from bore hole and hand dug well.

#### *3.1.8 Impairment of environmental quality*

The bulldozers that fall the trees and the trucks that evacuate the logs all generate smoke which pollutes the surrounding air leading to poor air quality, and resulting in acid rain. This leads to the impairment of the health of aquatic and terrestrial organisms around the reclaimed area.

#### *3.1.9 Improper disposal of solid waste*

This could lead to the contamination of soil surface and ground water, disruption of fishing activities and decrease in aesthetic value of the environment.

## 4. Impact of landscape reclamation on marine ecosystem

The marine coastal ecosystem is made up of different zones. The uppermost layer or photic zone is 10 m of water and absorbs 80% of solar energy, which carries out primary productivity. The warm shallow waters of the continental shelves are most biologically productive and support the greatest species diversity. Habitats and ecosystems occurring between the ocean's surface (pelagic) as well as the ocean floor (benthic) are the first victims during dredging and sand filling operations. This is because the organisms that reside in these areas are wiped out during the first phase of reclamation.

Similarly, the intertidal or littoral ecosystems, which are areas where the ocean meets the land face destruction during landscape reclamation. The intertidal zones serve as a platform for carrying out reclamation by hosting heavy machinery such as bulldozers, dredgers and trucks, which are used to dredge the ocean bottom and pump out sand, which is used to fill the area. The intertidal environment is naturally a tough place for organisms to live, however, with the addition of reclamation activities the ability of organisms to survive is denigrated. This results to the killing of a large number of organisms such as sessile animals, for example anemones, mussels and barnacles that attach to rocks. There are also some burrowing organisms that dig into shore sand. Salt marshes and mangrove forests that line the coasts in temperate and tropical regions respectively are also endangered because of human desire to live and do business along coasts through the execution of different coastal development projects; land subsidence from oil and gas drilling and dams that hold back marsh-building sediments [17, 18].

Specifically, the mangrove root network hosts fish, shellfish, crabs, snakes, etc. In the same vein, birds feed and nest in the dense foliage. Mangroves provide various ecosystem services such as food, medicine, tools and construction materials. Globally, half of the world mangroves have been destroyed [16] as people have reclaimed those areas as a result of coastal development. Shrimp farming in particular has resulted to the conversion of large areas of mangroves. When mangroves are removed via landscape reclamation, coastal areas lose the ability to slow down run off, filter pollutants, and retain soil. As a result, offshore systems such as coral reefs and eel grass beds are readily degraded. Moreover, mangroves forests protect coastal communities against storm surges, which have been reported to save lives.

## 5. The role of human ecology in land reclamation

People participation is very important in landscape reclamation because it would provide a consensus view on land use to the host community [19]. Five ways human ecology influence reclamation of coastal environment include:

1. *Concentration*: is a process whereby a given area becomes concentrated by human population due to the fertility of the land. Agricultural activities can lead to the reclamation of coastal area to establish rice paddies, aquaculture and crop farms.

2. *Centralization*: these are lands that are regarded as focal points as a result of their economic value. These areas dominate the hinter land and are often reclaimed and sand filled to expand their economic potential. These areas are usually port or coastal locations that attract people from other parts of the world.

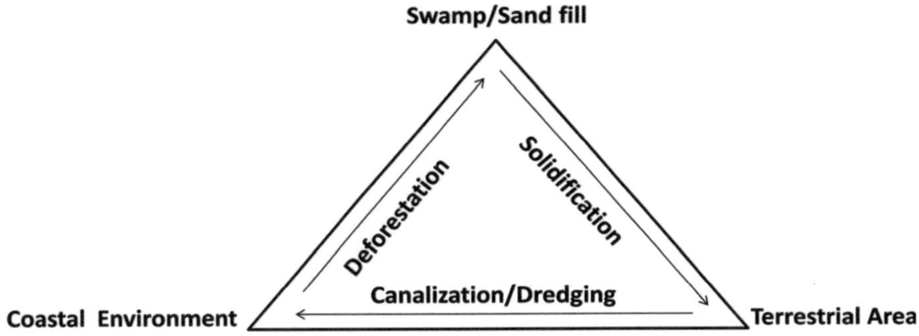


**Figure 7.**
*Conversion of coastal environment to terrestrial area through deforestation and solidification to terrestrial environment in the Niger Delta, Nigeria.*

3. *Segregation*: this is when units or communities cluster together based on homogeneity of ideals or goals. Homogeneous ideals such as customs, educational goals, and occupational goals all make up segregation. For instance, establishment of refinery in a locality leads to the reclamation of more coastal lands to harbor the cluster of people who come to seek for jobs.

4. *Invasion*: this is as a result of the migration of people into coastal communities in search of new technologies that provide job opportunities. An example is the establishment of the liquefied natural gas (LNG) plant in Bonny Island in the Niger Delta region of Nigeria. This facility is situated at a coastal community that is land-locked. Thus in other to expand the land area to accommodate the industrial complex, surrounding creeks were mowed down; sand filled and solidifies to build offices and apartments for workers.

5. *Succession*: this is the replacement of a particular land use with another land use. The conversion of a coastal area to a terrestrial area via developmental projects is a human-mediated succession. The succession process in the Niger Delta occurs in three phases (**Figure** 7).

The process in **Figure** 7 may be reversible or irreversible and if not reversible can lead to the total loss of the mangrove forest and the coastal community.

## 6. Restoration and management methods

Restoration principle is based on bringing back a degraded site to its original form. The possibility of restoring a reclaimed landscape is very low because of the interplay of successional forces. The conversion of a coastal environment to a terrestrial environment is absolute and cannot be reversed. Therefore, the best option is to remediate it so that it will carry out its ecological functions even if it results to another landform. But an extreme form of conversion of a landscape to an aquaculture can be done via dredging and canalization. This means areas that had been cleared and sand filled could be opened up again to be interconnected with the river, so that it will gradually become alive and revert to its original form [11]. This will however, take 10—20 years to stabilize. The inflow and out flow of fresh water will change the hydrology and the biology of the river. The area can be strictly protected against further anthropogenic activities to allow for evolutionary forces to change the restored environment. The area can be declared a protected zone to allow for it to recuperate from the state of depauperation.

*DOI: http://dx.doi.org/10.5772/intechopen.82053*

### 6.1 Case study: seedling recruitment experiment

Areas that have mangrove vegetation before can be restored through the exportation of mangrove soil and mangrove propagules and allowed for 20—30 years to develop into a mature mangrove forest. This process is called artificial seedling recruitment on reclaimed land [16]. In a natural seedling recruitment, after a disturbance event the first set of species that settle and colonize an area are the pioneer species. They gradually occupy the area through seedling recruitment process. In a classical case of land recuperation after landscape reclamation at Eagle Island in the Niger Delta, recruitment occurred through natural process within a space of 1–3 years. In 2014, an area measuring 100 m × 50 m was dredged and sand filled. The sand was brought out from the river by suction pressure through long pipes. The sand filled area became the dumping ground for sand, where trucks evacuated the sand to buyers. The side of the sand filled area was piled up to form balkanization against the inflow of river water. The sand mining activity was abandoned after 2 years of operation. In the course of this period a small outlet was created by the side of the sand filled area, which allowed the entry of river water into the sand filled area during high tide. Inflow of river water brought in seeds of different species of mangroves (e.g. red, white and black mangroves) and seeds of nipa palm and *Heritiera littoralis*. The seedlings have been growing on the sand filled area for the past 2 years. The plants growing on the sand fill area are between 0.5 m and 1.0 m tall. A field observation made indicates that seeds at the end of the sand filled area had better growth than seeds at the mouth of the entrance of the balkanized sand. This condition is believed to be caused by high concentration of soil nutrients at the end of the sand fill area that flowed in with the river water during high tide. Growth may also be facilitated by the absorption of soil nutrients embedded in the first layer of soil brought in by tides. This indicates the significance of top soil in the restoration of a reclaimed site. This is a natural ecological restoration. It is a regular progress or change by plant and animal due to natural or anthropogenic disturbances. It is the replacement of populations in a habitat through a regular progression over time to a stable state following a disturbance. In a preliminary study conducted in the sand dredged area, soil samples were collected for physico-chemical analysis at three sites (T1, T2 and T3) from the back of the sand filled area to the entrance of the sand fill area (**Figure 8**). Furthermore, census of species found in the area was made at the three study sites. The results of the physico-chemical analysis and the species abundance test were derived as shown in **Figure 9**. The result indicates that Potassium and magnesium had the highest concentration. Although they were not significantly different from each site ($F_{2,\ 12}$ = 0.19, P = 0.83). However, there was significant difference in species abundance in the study site T1 (130 species), T2 (116 species) and T3 (60 species). The most dominant species found was white mangroves (*Laguncularia racemosa*) (108) followed by red mangroves (*Rhizophora racemosa*) (104), black mangroves (*Avicennia germinans*) (77) and nipa palm (*Nypa fruticans*) (17). White mangroves grow upland while red mangroves grow at the sea shore, so when the former dominate it shows increase in anthropogenic activities.

The next kind of management is human management [19]. It involves human beings because they are the ones that cause problems for biodiversity. It is a system where plants and animals have advantage in reserved area. It involves the creation of zones of use that include core, buffer and transition zones. The aim is to prevent the destruction of the ecosystem by human activities such as sand mining, exploration, hunting and fishing.

**Figure 8.**
*Experimental design of recruitment experiment in a sand dredged mangrove forest at Eagle Island in the Niger Delta, Nigeria. Sites T1, T2 and T3 are the end, middle and mouth of the sand fill area.*

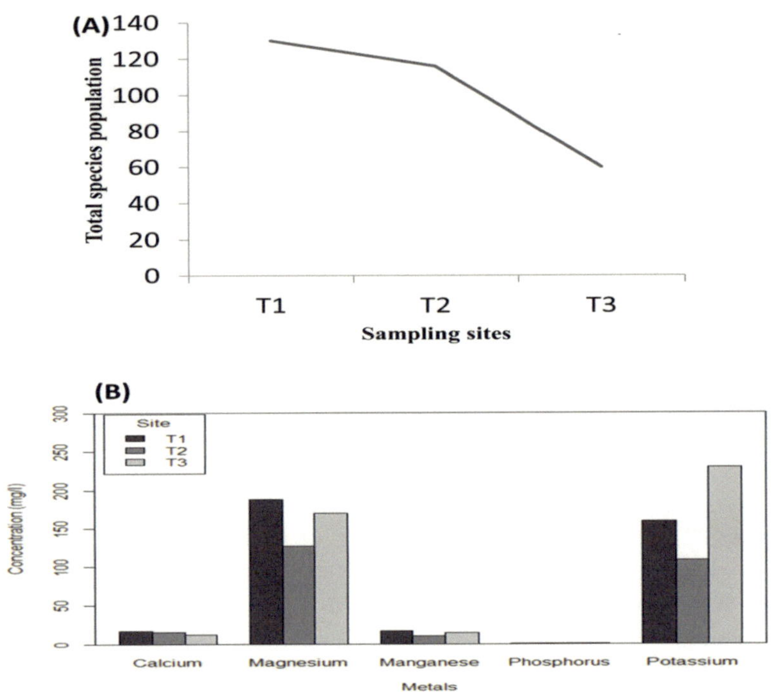


**Figure 9.**
*(A) Graph of species population along sampling points (T1, T2 and T3) indicates a decrease in number of species from point T1 (high nutrient content) to point T3 (low nutrient content); and (B) graph of concentration of physico-chemicals along within sites T1, T2 and T3.*

## 7. Ecosystem management

Together the system of management is called ecosystem management, which is a way of managing reserve to benefit biodiversity and people. It is a strategy for protecting or restoring the function, structure and species composition of an ecosystem while providing for its sustainable socioeconomic use. The method is a natural recovery and a passive method of accomplishing restoration. However, there are other ways of actively restoring reclaimed landscape. They include:

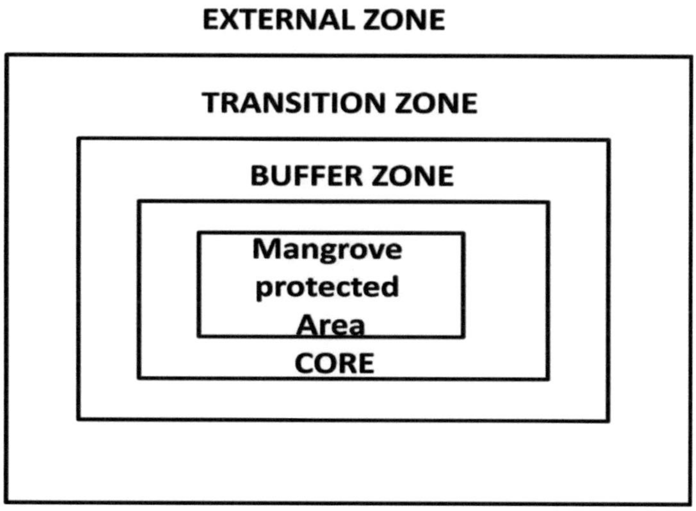


**Figure 10.**
*A proposed mangrove protect ted area design where no exploration, exploitation or reclamation activity will occur in the Niger Delta, Nigeria.*

1. *Replacement*: instead of going back to the original, which is impossible, it involves taking it to a different direction to create a replacement. We basically try to establish new habitat type because we cannot establish historic ecosystem. For instance, oil field that is established in mangrove swamp. Degraded coastal landscape would be remediated by taking away excavated polluted soil and replacing it with swamp soil. The water channel has to be set up to facilitate interconnectivity. The restoration of tidal force, edaphic factors and climatic effect will encourage natural recruitment of seedlings. The colonization of pioneer species will accelerate the establishment of other successional stages (e.g. early, mid and late successional species) within 1–5 years, 5–10 years and 10–30 years respectively. Another example is the replacement of invasive nipa palm forest with native mangrove forests.

2. *Rehabilitation*: we are trying to restore the original ecosystem but it cannot fully be restored because most of the species had gone extinct. An example is a construction project such as the installation of infrastructure on reclaimed coastal areas e.g. sea-side resort, roads, houses, light poles and shopping malls.

3. *Restoration*: it is the attempt to fully restore the original ecosystem. An example is a nipa palm invaded area. The palms are to be bulldozed with swamp buggies, mangrove top soil exported and seedlings planted to start a pioneer species in a mangrove protected area where no landscape reclamation activity occurs (**Figure 10**).

## 8. Conclusion

Landscape reclamation is used to remediate polluted and devastated area in many parts of the world such as abandoned coal mine, crude oil exploration site and abandoned aquaculture. But in the Niger Delta landscape reclamation is used to convert coastal areas to terrestrial areas. Two major activities that devastate coastal areas are sand mining and off-shore dredging. Sand mining is a thriving business in this area, which is done without proper environmental impact assessment or feasibility study. Continuous mining leads to the deformation of the coast lines and the destruction of aquatic organisms, which results in extinction of species. Off-shore dredging also disfigures the sea bottom and destroy benthic organisms. Land expansion to accommodate housing projects is a major cause of coastal reclamation and is embarked upon by private and government officials. The suggested solution to revert an already devastated area is by applying natural and human mediated ecological principles to facilitate land and coastal recovery [20].

## Author details

Aroloye O. Numbere
Department of Animal and Environmental Biology, University of Port Harcourt, Choba, Nigeria

*Address all correspondence to: aroloyen@yahoo.com

## References

[1] Macnaghten P. Theory, culture and society. Nature. 2006;**23**:347-349

[2] Birol E, Karousakis K, Koundouri P. Using economic valuation techniques to inform water resources management. A survey and critical appraisal of available techniques and an application. Science of the Total Environment. 2006;**365**: 105-122

[3] Numbere AO. The impact of oil and gas exploration: Invasive nypa palm species and urbanization on mangroves in the Niger River Delta, Nigeria. In: Makowski C, Finkl C, editors. Threats to Mangrove Forests. Coastal Research Library. Vol. 25. Cham: Springer; 2018

[4] Brooks CN. A model for redeveloping complex, highly contaminated sites in the Industri-plex site in Woburn, Massachusetts. WIT Transactions on Ecology and Environment. 2006;**94**:229-238

[5] Panagopoulos T. Linking forestry, sustainability and aesthetics. Ecological Economics. 2009;**68**:2485-2489

[6] Loures L, Burley J, Panagopoulos T. Post industrial landscape redevelopment: Addressing the past envisioning the future. International Journal of Energy and Environment. 2011;**5**(5):714-724

[7] Manzoor A, Hussain N, Niazi BH. Use of chemical amendments for reclamation of saline-sodic soils. International Journal of Agriculture and Biology. 2001;**3**(3):305-307

[8] Panagopoulos T. Reclamation of sites mined for lignite in northern Greece. In: Fantechi R, Balabanis PP, Rubio JL, editors. Desertification in European Context: Physical and Social Economic Aspects. Belgium: European Commission Brussels; 1995. pp. 575-582

[9] ESRI. Arc GIS 9.1. Redlands, CA, USA: Environmental Systems Research Institute; 2006

[10] Kathiresan K, Bingham BL. Biology of mangrove ecosystems. Advances in Marine Biology. 2001;**40**:81-251

[11] Punter J. The welsh development agency design guide. In: Its Role in Raising Standards in Wales. Cardiff: Welsh Development Agency; 2002

[12] CEDA. Coastal Profile of Nigeria. Abuja: Federal Environmental Protection Agency; 1997

[13] Numbere AO. Impact of invasive nypa palm (*Nypa fruticans*) on mangrove forest in the Niger Delta. In: Makowski C, Finkl C, editors. Coast in Crisis. Coastal Research Library. Vol. 28. Cham: Springer; 2018

[14] De Sousa CA. Turning brown fields into green space in the city of Toronto. Landscape and Urban Planning. 2003; **62**:181-198

[15] Numbere AO, Camilo GR. Mangrove leaf litter decomposition under mangrove forest stands with different levels of pollution in the Niger River Delta, Nigeria. African Journal of Ecology. 2017;**55**(2):162-167

[16] Alongi DM. Mangrove forest: Resilience, protection from tsunamis, and responses to global climate change. Estuarine, Coast and Shelf Science. 2008;**76**(1):1-13

[17] Pinno BD, Sherr I, Errington RC, Shea K. Islands-soil patches and plant community dynamics on a news oil sands reclamation design. Journal of American Society of Mining and Reclamation. 2016;**5**(1):28-44

[18] Loures L, Panagopoulos T. Reclamation of derelict industrial land in Portugal: Greening is not enough. International Journal of Sustainable Development and Planning. 2010;5(4): 343-350

[19] Loures L, Crawford P. Finding public consensus: The relevance of public participation in post-industrial landscape reclamation. In: 1st WSEAS International Conference on Landscape Architecture (LA'08); Agarve, Portugal. 2008

[20] Ayala R, Ramirez J, Camargo S. Volaracion de la calidad fragilidad visual del paisage en el valle de zapotitlan de las Salinas, Puebla (Mexico). Spain: Faculdade de Geografia e Historia da Universidade de Madrid; 2003

Chapter 5

# Public Perceptions of Values Associated with Wildfire Protection at the Wildland-Urban Interface: A Synthesis of National Findings

*Jason Gordon, Adam S. Willcox, A.E. Luloff, James C. Finley and Donald G. Hodges*

## Abstract

The wildland-urban interface (WUI) continues to transform rural landscapes as previously undeveloped areas are populated with residential and commercial structures which, in turn, impact ecosystems and create landscapes of risk. Within this context, the science of wildfire risk mitigation has experienced renewed and enhanced support among scientists and managers. However, risk mitigation measures have not found purchase in either the public's acceptance or involvement in this new role of and for fire. This may partially result from little regard for the effects of wildfire prevention efforts on values other than protecting homes and other structures. We report findings from qualitative interviews conducted across the United States to identify and define various values at risk from wildfire. Values influencing risk mitigation emerged from the biophysical, sociodemographic, and sociocultural contexts of wildfire. Findings demonstrate how wildfire is intertwined with diverse sets of risks experienced in daily life. We provide a discussion of how this research impacts the transformation of landscapes and risk management strategies. Identifying and better understanding the effects of values associated with wildfire—and landscape change in the WUI—will allow natural resource managers and decision makers to develop more effective fuel treatment programs and land use policies.

**Keywords:** wildfire, wildland-urban interface, public perception

## 1. Introduction

For generations, the public was told that fire destroys forests and many of its associated values (e.g., timber, wildlife, recreation, esthetics, ecosystem services). Recently, the science of fire prevention and fuel treatments has experienced renewed and enhanced support particularly as resource managers have learned more about ecosystems, their functions, and feedback loops. Still, wildfire prevention measures for enhancing ecosystem services have not found purchase in either the public's acceptance or involvement in this new role of and for fire.

This is especially true of wildland-urban interface (WUI), a landscape of transition whereby increasing numbers of people and built structures invade wildlands. In the WUI of the United States, fire protection is directed not only at forests but also at homes and structures that are much more prevalent there. Resistance to recommended fuel treatments arises from two primary factors: (1) many of the prescribed fuel treatments do not reflect residents' understanding of forest management and (2) treatments are developed with little recognition of the multiple values owners and the public place on forests [1–3].

A limited number of studies have examined the relationship between wildfire mitigation activities and amenity values, recreation, or sense of place (e.g., [4–8]). More research is needed to consider the full set of multiple and competing values, particularly because wildland fire policy has evolved from agency-focused risk mitigation to empowerment and action at the household and community levels [9, 10]. Risk managers must acknowledge that successful implementation of risk reduction strategies necessitates resident participation which, in turn, demands an understanding of values associated with wildfire protection.

This chapter synthesizes findings from research exploring wildfire risk perceptions as they reflect variation in social values of the forest surrounding the study communities. To do this, we present findings from key informant interviews, which comprised the first phase of a multi-phase, mixed methods project that also included facilitated community discussions and a national mail survey. Findings will help fire managers, community leaders, and other end-users better understand public perceptions of issues surrounding the full range of values associated with rural-urban transition zones. The definition of "public" is as dynamic as the forest, and it is land managers' responsibility to recognize public concerns and tailor their messages and activities to them.

## 2. Literature

### 2.1 Risk perceptions of wildfire

Risk perceptions of wildfire have reflected the variability of findings found in the broader risk perception literature. Despite the importance of risk perceptions to wildfire mitigation actions [11–16], there is little consistency in the literature regarding the relationship between risk perceptions and mitigation behavior. For example, previous experience has been a factor in creating defensible space around homes; as well, it has been associated with apathy regarding the perceived likelihood of repeated wildfire events [17]. Similarly, proximity to wildfire has been shown to increase concern [18] or have little significance [19]. Inconsistencies suggest intervening social, economic, and ecological elements influence the ways homeowners view and address a wildfire hazard [20, 21]. As noted by several authors, risk perception is important to mitigation actions, but other factors within the biophysical, sociodemographic, and sociocultural contexts of wildfire may play equal or more important roles [22, 23].

### 2.2 The biophysical context

**Figure 1** depicts a matrix of biophysical, sociodemographic, and sociocultural contexts affecting landscape change and risk perceptions. The matrix reflects the idea of ecological roots, or the connection between ecological, social, economic,

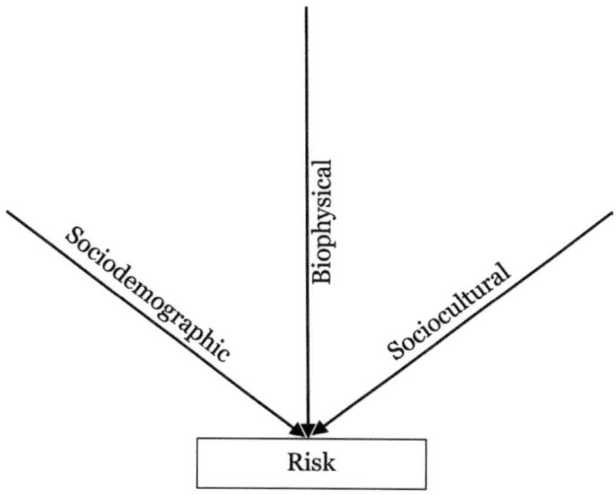


**Figure 1.**
*Biophysical, sociodemographic, and sociocultural contexts (modified from [24]).*

and cultural resources [25]. Ecological roots encourage landscape recovery by finding solutions linking these four elements.

Biophysical factors include land use and cover, topography, climate, fuel load, fire regimes, wildlife, and numerous other ecological characteristics of the study site. An important consideration in the WUI is that wildfire risk can increase due to forest harvesting and fragmentation, inappropriate landscaping decisions, and flammable home building materials [26–29]. Landscape type, that is, shrub land versus pine forest or upland hardwood forests, also responds differently to fire susceptibility [30]. Emergency responders and natural resource managers are obligated to protect life and property even when low-density housing is difficult to defend from wildfire due to long emergency response times, lack of water sources, and underdeveloped road systems [31].

Recent studies have found WUI residents were aware that the lack of fire led to unhealthy forest conditions [21, 32, 33]. As well, public education and outreach has had a positive influence on knowledge about fuel reduction treatment methods [34, 35]. Some studies have found respondents accepted the use of prescribed burning as an ecological practice on public land, although they preferred mechanical treatments when the forest is located close to a populated area [32, 36, 37]. Despite these findings, high knowledge levels have also been associated with decreased concerns [20, 38]. Ample evidence that the public recognizes the ecological role of fire suggests that additional factors intervene in attitudes toward fuel treatment methods, decisions to live in high-risk places, and other factors related to wildfire risk perceptions and mitigation activities.

### 2.3 The sociodemographic context

Transitional landscapes are defined not only by biological characteristics, but also by social changes, which should be considered in understanding risk perceptions and mitigation. For example, Moreira et al. [39] literature review noted socioeconomic drivers favoring land cover changes, which, in turn, contributed to increasing wildfire risk. Still, although some differences have been observed for race and gender, sociodemographic variables by themselves have rarely explained important differences in attitudes and behaviors toward wildfire [3, 7, 15, 17, 40, 41]. This includes urban or rural residency status, which has been linked to divergent views in resource management (i.e., prescribed fire, thinning, grazing), but not significant differences in wildfire risk perceptions per se [41]. Still, sociodemographics are important in the WUI where social and landscape change can lead to conflict in the sociocultural context of wildfire [42]. Surveys used to quantify perceptions of wildfire risk often have a limited ability to draw connections between sociodemographic variables and more difficult to measure sociocultural variables, including social conflict and collective agency [43]. Improving social assessment related to land use change is critical given that land use policies and land cover management affect fire patterns and intensity [39].

### 2.4 The sociocultural context

Traditions, attitudes, beliefs, and value systems of WUI populations are part of the sociocultural context of wildfire. As Beck [44] noted, risk is intertwined with society's economic and political structures, and the complexity of modern risk means no one fully understands the dangers they face. Wildfire risk is a fusion of ecological and technological drivers complicated further by enormous budgets, political lobbying, and media amplification. Nevertheless, risks are bound up in instrumental rational control—through decisions people make about their lives and future courses of action [44]. The sociocultural context of risk underscores both agency and the social construction of wildfire and preparedness (also [45]).

Within the sociocultural context, in-migration is an important characteristic of many interface zones even though the WUI is not defined by population change. In such cases, residents often have to renegotiate local identity and symbols of collective life as the penetration of new and different value systems and threats to traditional norms and membership groups occurs with the emergence of heterogeneity [42, 46]. These differences can manifest in competing notions about esthetics, land use and community growth, natural resources management, and strategies for addressing wildfire risk. Additional community processes (e.g., race and class conflict, environmental concerns, historical grudges, social movements, cultural celebrations) emerge from change. Sharing direct and indirect wildfire experiences and local landscape knowledge can become increasingly challenging as a result of conflicting worldviews and competing group memberships [22, 47, 48].

Social institutions (i.e., insurance, government, and corporations) are a critical aspect of the sociocultural context because they are key players in the management of risk [44]. To maintain this role, effective messaging must be continuously reevaluated because trust is dynamic and highly dependent on the relationship between the institution and the public. Residents who have less confidence in an institution's ability to fulfill its role have demonstrated increased levels of concern about wildfire and are less likely than others to accept initiatives designed to address it [21, 29]. Risk communication through direct contact, citizen involvement in decision-making, and an understanding of local context has been most effective in sustained wildfire mitigation at the individual and community levels [16, 27, 34].

To examine values associated with wildfire perceptions and risk management, we consider the three dimensions of wildfire risk outlined above: biophysical, sociodemographic, and sociocultural. We then describe how community mitigation actions materialized from the contexts and merged into effective risk reduction strategies. Findings can serve as a springboard for wildfire and fuel treatment message development and contribute to evaluation processes capable of use in areas where public education campaigns are ongoing, are required, or are appropriate.

## 3. Study area

Research sites for this study were selected based on their classification in the WUI [31] and whether or not they had a Community Wildfire Protection Plan (CWPP; http://www.forestsandrangelands.gov/success/index.cfm). Sites varied by sociodemographic indicators (e.g., in-migration and seasonal housing) and biophysical factors (e.g., forest type) and corresponded to US Forest Service Regions (Eastern, Pacific Northwest, Southern; [49]). For comparison, a metropolitan county was selected with an adjacent nonmetropolitan county in each region. All sites had been designated as wildfire-prone and wildland fire issues were prominent [9]. Human-driven changes have had an important influence on land cover dynamics, including landscape diversity [50].

Study counties included Clinch (nonmetropolitan) and Lowndes Counties (metropolitan) in Georgia; Carlton (nonmetropolitan) and St. Louis (metropolitan) Counties in Minnesota; Rio Arriba (nonmetropolitan) and Santa Fe (metropolitan) Counties in New Mexico; and Jefferson (nonmetropolitan) and Deschutes (metropolitan) Counties in Oregon. In 2010, populations ranged from under 6798 in Clinch County to over 200,000 in St. Louis County. All counties except Clinch, St. Louis, and Rio Arriba experienced an increase in population between decennial censuses. Deschutes County experienced the greatest population increase (37% to 157,733); Clinch County the greatest decline (−1% to 6798). Jefferson County experienced the largest amount of in-migration to rural areas (17% to 13,710), while Rio Arriba County lost the most rural population (−2% to 40,246).

## 4. Methods

Key informant interviews were conducted in each of these four states during 2012 (**Table 1**). Interviews were administered to individuals knowledgeable about WUI issues and local affairs; moreover, these people were broadly representative of private forest landowners (PFLs), forest industry, government, local political and social factions, and social status [51]. We conducted 33 interviews in Georgia, 35 in New Mexico, 35 in Minnesota, and 45 in Oregon for a total of 148. Initial key informants were identified using local directories and Internet sources. To ensure

| | Number of key informants | | | |
|---|---|---|---|---|
| **Type** | **New Mexico** | **Georgia** | **Minnesota** | **Oregon** |
| Government (federal, state, local) | 13 | 23 | 16 | 28 |
| Business | 5 | 4 | 6 | 4 |
| Minority group | 2 | | 2 | 3 |
| Environmentalist | 4 | | 3 | 4 |
| Media | 2 | 2 | 3 | 2 |
| Landowners | 4 | | | |
| Forest industry | 1 | 3 | 3 | 2 |
| Citizen activists | 2 | | 1 | |
| Religious leaders | 2 | 1 | 1 | 2 |
| Total | 35 | 33 | 35 | 45 |

**Table 1.**
*Key informant types (n = 148).*

comparability across urban and rural counties of each state, informants representing each of the following perspectives were interviewed in each community: (1) federal and state land manager; (2) extension agent; (3) local planner and/or natural resource manager; (4) emergency services professional; (5) elected official; (6) business leader; (7) landowner; (8) religious leader; (9) journalist; (10) consultant or industrial forester; (11) environmental activist; and (12) citizen activist. Additional informants were identified using snowball sampling with purposive selection to encourage diverse perspectives, including an underrepresented or marginalized segment of local society [52].

Each interview covered (1) awareness of past and proposed fuel treatments in the area; (2) range of values associated with the WUI; (3) perceptions of wildfire risk; (4) public response to wildfire risk and occurrence; and (5) constraints on implementing wildfire reduction treatments. Open-ended questions encouraged informants to volunteer information, rather than simply responded to queries. Their rich and spontaneous replies provided a reality view of a place, including broad relationship patterns among actions and actors with the local environment [53].

Interviewers' notes were analyzed for emergent themes using a two-step coding process involving reading the notes and then coding into thematic categories [54]. Themes were compared within and across cases in each state and then over the four states. Each author reviewed the data and added additional interpretation to improve reliability.

## 5. Findings

### 5.1 Risk perceptions

Risk perception themes were related to both wildfire and wildfire prevention with a focus on (1) wildfire concern and (2) smoke. Aside from Jefferson County, where communities along a main highway were surrounded by irrigated crops, residents from rural communities in the study area demonstrated awareness about wildfire. In highly wildfire-prone areas, long-term residents mentioned wildfire as a routine aspect of life, especially during "wildfire" season. For example, a Minnesota respondent remarked, "We were raised with that awareness. We know to be careful as individuals" (St. Louis County). As discussed further below, this attitude may be waning in communities characterized by in-migration of residents seeking recreation and retirement opportunities. In such cases, perceptions tended to be mixed, ranging from heightened concern to indifference.

Of the four study areas, wildfire was mentioned as a prominent risk worthy of immediate concern only in Deschutes County (although CWPPs existed in each site[1]). In the other sites, wildfire became a priority when residents perceived a near and eminent danger. A respondent in Lowndes County noted, "Fires in Colorado: we see and hear about [them] but that's just another news story until you smell and see the smoke in your neighborhood and communities." By contrast, previous experience with wildfire was mentioned as a source for influencing risk perceptions only when the disaster event was recent (within the last 10 years) and catastrophic.

As the Georgia participant noted, smoke increased concerns about a wildfire even if personal safety and property were not under immediate threat. In particular, rural

[1] Although CWPP requires resident input (HFRA 2003), most of our informants were unaware of local CWPPs. Because CWPP is a necessary condition for receiving financial aid for mitigation under the Healthy Forests Restoration Act (2003), the critical elements of participation may have been neglected during the CWPP process in order to receive funding.

residents were likely to mention the negative health effects from smoke linked to prescribed fuel reduction fires or controlled fires employed in agriculture field preparation. Some informants said official and unofficial communication about smoke effects had wider socioeconomic impacts, especially in areas reliant on outdoor recreation. For example, a rural Minnesota resident stated, "The Pagami Creek Fire did not put smoke into Ely. The headlines outside of the community were that the town was on fire. These headlines killed business for the season" (St. Louis County). In urban areas and communities frequented by seasonal residents and vacationers, smoke was perceived as a risk if it disrupted community events, personal recreation activities, and travel. Although natural resource and emergency managers acknowledged wildfire risk specifically, the broader population was often more concerned about smoke.

### 5.2 Biophysical context

The predominant themes from the biophysical context were (1) water availability; (2) proximity to fuels; and (3) fire as part of a natural system. Water was overwhelmingly mentioned as an issue in the Western states. For example, several New Mexico informants connected noticeable decreases in water availability with climate change and mountain pine beetle (MPB; *Dendroctonus ponderosae*). Wildfire was seen as a corollary to these hazards. In the quote below, an informant noted a decrease in precipitation with less snow hitting the forest floor due to a thick canopy.

> *The snow pack is no good - it does not hit the ground because the trees are too thick. We are not getting runoff like before, so it's too dry. MPB is threatening stands in NM...There is too much fuel loading and fire suppression. (Rio Arriba County)*

This description suggests that high tree density increases wildfire risk because snow caught in the canopy sublimates.

Findings related to the biophysical context overlapped with the other dimensions of wildfire. For instance, Western residents connected water issues with sociodemographic themes such as population growth. Water problems were perceived as worsening with the growing WUI, but policy had not kept pace with water use. In some cases, study participants made a connection between management of local watersheds and wildfire. More importantly, WUI growth was a common denominator for both water use and wildfire.

By comparison, proximity to wildfire fuels emerged most prominently in Oregon and Minnesota. These places also exhibited the greatest degree of knowledge about fire's role in forest ecosystems. Minnesota informants noted that wildfire hazard increased proximal to a designated wilderness area. There, a catastrophic storm toppled extensive timber volume in the late 1990s. Through the media, word of mouth, and recreation activities, residents were frequently reminded of the well-known blowdown: "Fire is simply not thought about very often [in Carlton County]. However, the Boundary Waters incident makes people think about what could happen."

Similarly, in Oregon, Jefferson County informants noted forests were not near their agricultural communities. However, Deschutes County residents were in closer contact with the forest and concerns increased with knowledge of forest conditions. Informant responses highlighted how wildfire risk perceptions varied across minor geographic areas in relation to biophysical characteristics.

Perceptions of fire as part of an ecological system differed within and across study sites. Although such attitudes can reflect biophysical conditions, they were often intertwined with sociocultural values. Some, as in rural Georgia, considered idle land the antithesis of nature. Plantation forests were as natural as nonplantation forests, and human intervention, including prescribed burning, in the forest was natural. "(The) difference between a planted forest and natural is not much... Controlled burning is insurance against wildfire – fire is our best friend" (Clinch County). By contrast, rural Minnesota informants often thought that because their forests were "overmanaged," they were not natural and prescribed burning was an artificial event (although not necessarily an unwanted practice). Rural Oregon informants said contemporary forests were not natural, but prescribed fire simulated a natural event to improve forest health.

### 5.3 Sociodemographic context

There were two predominant themes in the sociodemographic context: (1) population change and (2) economic shifts. In all states, residents discussed population change. Depending on the site, change was linked to WUI encroachment into fuel zones, increased diversity, and loss of community identity. In Georgia, New Mexico, and Oregon (Jefferson County), racial conflict emerged as an aspect of population change. Oregon and New Mexico were experiencing increasing immigrant populations, while the Georgia sites were characterized by race-based residence patterns. As a result, the sociodemographic context tested local social relationships and capacities to agree on values associated with fire and fuels management. For example, new and seasonal residents were said to be less concerned with community wildfire hazard impacts, but very concerned with smoke. Permanent residents, by contrast, either expressed concern or were not concerned about both aspects of fire.

Overall, Oregon communities sharply contrasted with the other sites in terms of dealing with different population groups. Despite responses such as "The term local ... is a misnomer because of the vast growth that this area has experienced" and "It is hard to do things when people are not the same," the communities were described as proactive regarding wildfire preparedness. A local NGO specifically dealing with coordinating wildfire efforts was largely credited with leading community-wide efforts.

> *We were on the forefront of developing fuel management programs...Even the large landowners have gone on to do mitigation work without any assistance because it is the right thing to do. The program has grown in an organic way – it has not been forced.*

In New Mexico and Oregon, diversified and progressive economies of the metropolitan areas contrasted with the comparison rural agricultural counties. Economic strength was linked with perceived high levels of human capital, which in turn translated to successful wildfire preparedness strategies, among other collective concerns. Santa Fe County leaders (private and public) engaged in fuel reduction to protect its watershed. Deschutes County leadership, led by the NGO mentioned above, engaged residents in defensible space, collaborated with the U.S. Forest Service, organized homeowner fuel removal and disposal events, and constructed a FireFree demonstration home, among other activities.

In all sites, informants expressed concerns about rural poverty and the decline of traditional agricultural economies and population. This quote from St. Louis County illustrates emergent conflict as communities struggled to reconcile extraction-oriented and recreation-oriented values. Long-time, permanent residents saw the recent influx of seasonal residents and tourists as "ideologues ... [newcomers] like the appearance of the town, and then they are shocked there is support for mining. It is not that [permanent residents] support [mining], but that they want to see it be done to support the economy and their livelihoods." Notably, rural residents were more accepting of timber management; however, they were just as unlikely as their urban counterparts to discuss involvement in wildfire mitigation activities.

The sociodemographic context helped frame sociocultural values associated with wildfire preparedness and fuel reduction.

### 5.4 Sociocultural context

The main themes from the sociocultural context revolved around (1) natural resource values; (2) trust in government; and (3) community participation. Urban informants often discussed how residents valued forests for recreational pursuits (the main exception was rural St. Louis County where both urban and rural residents discussed recreation). In all sites, rural residents focused on cultural values associated with agriculture. Major conflicts over natural resource values often converged on motorized versus nonmotorized recreation (e.g., Minnesota), resource utilization versus ecocentrism (e.g., New Mexico), and progrowth versus restrained growth into wildlands (e.g., Georgia). Such conflicts often overshadowed collective definitions of risk and acceptable mitigation strategies. For example, in New Mexico and Oregon, environmental activist groups from urban areas were seen as opposed to management actions that included thinning, prescribed fire, and postwildfire salvage logging. This quote illustrates nuances in disagreements over management philosophies:

> *There are some environmentalists that say just leave it alone [and not to allow fuel reduction treatments]. But no one here wants to just clearcut the forest, they want responsible management. (Santa Fe County)*

Trusting the government to fight wildfires, conducting prevention treatments, and working with communities differed among and within study sites and related to past interactions and experiences with government agencies. Although interviewees indicated a degree of trust in the ability of state and municipal governments to fight wildfires, they expressed less confidence in the federal government to implement wildfire prevention treatments and programs. All of our study sites, except Georgia, had reportedly experienced escaped prescribed fires, leading to skepticism regarding treatment needs and agencies' ability to implement prescriptions. This quote from Oregon exemplifies the difficulty of balancing public acceptance of fuel reduction treatments.

> *If we do a prescribed fire and unexpected weather conditions crop up do we [local government] and forest service know what we are doing? This is a concern and is problematic. The pendulum that swings back and forth all the time – too aggressive logging the forests and now too aggressive burning them down – still sorting out how to balance this out. (Deschutes County)*

All study sites reported positive community response and increased volunteerism when faced with natural disasters, including wildfires. A high degree of participation in daily activities was described; informants said residents were proud of their communities and enjoyed contributing to improved local well-being in various ways. Like many communities, they also experienced major divisions along racial, ethnic, and socioeconomic lines. Participation was often linked to group membership. For example, a New Mexico informant described wildfire preparedness activities, including fuel reduction demonstrations; however, the Hispanic population was not involved and there was little expectation they would participate. One informant noted a constant question among residents was "Whose town is it anyway?" Informants from Minnesota, New Mexico, and Oregon described contradictory values and resources among social groups, which exponentially impacted conflict regarding other local life issues as well as disparate effectiveness of wildfire preparedness messaging between groups.

## 6. Discussion

This chapter presented a synthesis of key informant findings from four states regarding public perceptions of values associated with wildfire protection. Findings provide a foundation for future research and application by placing public wildfire risk perceptions into the biophysical, sociodemographic, and sociocultural contexts of communities, which frame the full range of values, attitudes, and behaviors associated with wildfire and forests (**Figure 2**). The model is fluid, with each dimension influencing and influenced by the other dimensions. Findings underscore the need to better measure and understand how sociocultural factors are associated with wildfire response.

Most key informants described fairly low community level concerns about wildfire despite their counties being recognized as high risk and having Community Wildfire Protection Plans in place [9]. Deschutes County was an exception, with community-wide risk mitigation actions reflecting relatively high levels of concern. Interviews from the two Western states illustrated residents' understanding of additional risks associated with fuel regimes. In some cases, informants were concerned about increased potential for wildfire as a result of tree mortality caused by the mountain pine beetle. In other cases, wildfire was linked to drought, which, in turn, was linked to climate change. Key informants noted population pressures can increase this web of concerns in the American West. These findings underscore the public's capacity for perceiving associations between environmental vulnerabilities; however, social and cultural dimensions of the community may cloud interpretations, and residents may have difficulty articulating such relationships.

Related to the sociodemographic dimension, risk perceptions, concerns, and behaviors varied across geographic areas in relation to biophysical characteristics of the landscape and cultural values of the population. This finding contrasts with previous research failing to demonstrate sociodemographic differences, including urban or rural residency [3, 17, 55]. Our study found increased concerns in the West compared to the two other study areas primarily due to the regions' recent history of catastrophic wildfires and drought conditions. As well, prescribed burning was generally accepted, but perspectives varied regionally as to whether it mimicked natural processes or was a component of "unnatural forests." In addition, findings generally supported previous research demonstrating public understanding of

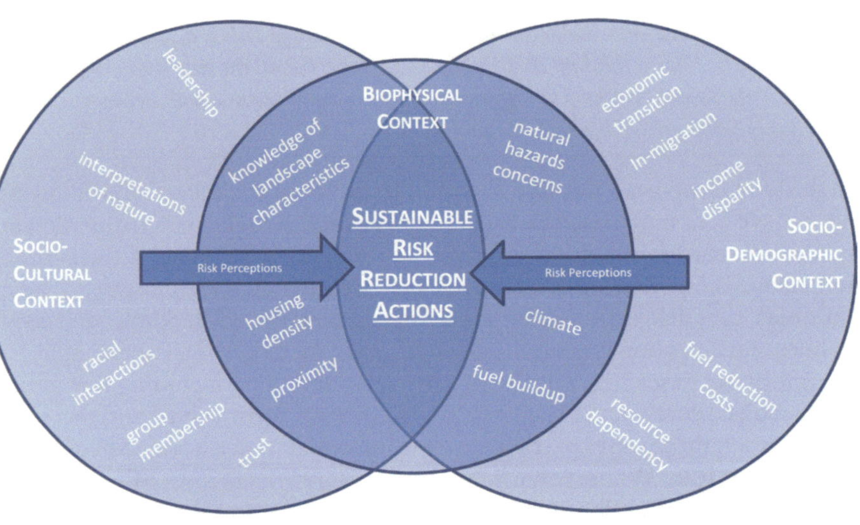


**Figure 2.**
*Framework for understanding sustainable risk reduction.*

wildfire in forest ecosystems (e.g., [33, 45, 56–58]). However, rural residents, who informants described as having more knowledge of biophysical processes than urban residents, were typically less concerned than their counterparts.

This study acknowledges that differences in wildfire perceptions are entwined in society's sociocultural structures, which evolve along with the changing ecology of transitional landscapes. As McCaffrey and Olsen [43] noted, sociocultural factors are "the more complex, often identity-based, and harder-to-measure factors ... [that] appear more likely to explain variation in how individuals respond to fire management issues." Key informant interviews demonstrated that growing communities have substantial challenges related to the social milieu, which influenced collective perceptions and the ways communities could address wildfire [22, 42, 47, 48]. The importance of sociocultural factors suggests a considerable need for work that improves the ability to identify and describe how the sociocultural context fits into risk perceptions and mitigation, key components of managing transitional landscapes, along with ecological restoration and environmental activities.

As well, our findings underscore the importance of local social and political institutions as key players in leading risk management [44]. Of our eight county sites, Santa Fe and Deschutes Counties demonstrated the strongest actions to reduce wildfire hazard. These actions were driven by robust collaboration between public and private groups (also [59]). Organizations, such as the NGO in Deschutes County, were important for transcending sociocultural divisions within communities and initiating dialog about the threat of wildfire to *all* residents. Although New Mexico exhibited mitigation activities, key informants acknowledged the activities excluded specific segments of the population; therefore, the actions were impeded at the community level.

Several implications emerge from this discussion. Because residents cope with a range of risks and hazards in their daily lives, many of which emerge from the changing social and biophysical landscape of the WUI, risk managers and community leaders should acknowledge competing risks when developing wildfire mitigation programs and messages. Risk managers have long understood that individuals have a finite capacity to effectively address many risks they face, and wildfire may take a backseat to other more salient concerns. This study suggests one way of initially identifying and characterizing competing risks and concerns is to employ the framework used here for understanding sustainable risk reduction (**Figure 2**). With residents' direct input, this approach can help promote dialog, understanding, and prioritization of community level concerns. As evidenced by Deschutes County leadership, resident contribution to the classification of risks is critical to the success of a comprehensive framework.

In addition, findings from this study suggest the potential for increasing awareness about wildfire and reducing risk by incorporating intersecting concerns from the biophysical context and acknowledging geographic differences [7]. For example, because water concerns and mountain pine beetle were salient and linked to wildfire, such hazards could be used as a vehicle to develop messaging that specifically addresses wildfire mitigation activities at the individual and community levels. For many places in the arid West, fuel reduction within the immediate goal of water quantity (and watershed management) may be valued more than fuel reduction for its own sake.

Similarly, programs in Georgia might focus more on the local health effects of smoke produced by wildfires; forest management activities such as prescribed fire have the potential to reduce airborne contaminants. One goal of this communication would be to allay the remnants of the total suppression message. Because residents are familiar with smoke, but not familiar with wildfire per se, communication strategies addressing smoke may have the desired impact. This implication may have ancillary effects on the scale of fuel reduction since all prescriptions must be implemented on a larger scale than currently performed in order to have any real difference in mitigating wildfire risk [60].

Finally, strong leadership can drive the process of characterizing community members' diverse values and concerns. Ideally, leadership in wildfire risk mitigation would originate from a community organization with no political affiliation. It is critical that, as one Oregonian put it, the program grow "in an organic way" and be part of the broader community development process [38, 61]. Although leadership has been a factor in the human dimensions of wildfire literature [7, 41, 62, 63], further research is needed to characterize the qualities and processes (e.g., dispute resolution, social learning, and collaborative planning, which require an understanding of the ecological and social dynamics of the locality) of leadership emerging in success stories [43, 59, 64, 65].

## Acknowledgements

Funding for this project was provided by the Joint Fire Science Program Project ID 11-1-9-12.

## Author details

Jason Gordon[1*], Adam S. Willcox[2], A.E. Luloff[3], James C. Finley[3] and Donald G. Hodges[2]

1 Mississippi State University, Starkville, United States of America

2 University of Tennessee, Knoxville, United States of America

3 The Pennsylvania State University, United States of America

*Address all correspondence to: jg966@msstate.edu

## References

[1] Finney MA. The challenge of quantitative risk analysis for wildland fire. Forest Ecology and Management. 2005;**21**:97-108

[2] Stein SM, Menakis J, Carr MA, Comas SJ, Stewart SI, Cleveland H, et al. Wildfire, Wildlands, and People: Understanding and Preparing for Wildfire in the Wildland-Urban Interface—A Forests on the Edge Report. USDA Forest Service, Rocky Mountain Research Station Gen. Tech. Rep. RMRS-GTR-299. Fort Collins, CO; 2013

[3] Toman E, Stidham M, McCaffrey S, Shindler B. Social Science at the Wildland-Urban Interface: A Compendium of Research Results to Create Fire-Adapted Communities. U.S. Department of Agriculture, Forest Service, Northern Research Station Gen. Tech. Rep. NRS-111. Newtown Square, PA; 2013

[4] Brenkert-Smith H. The place of fire. Natural Hazards Review. 2006;**7**:105-113

[5] Collins T, Bolin B. Situating hazard vulnerability: people's negotiations with wildfire environments in the US Southwest. Environmental Management. 2009;**44**:441-455

[6] Gunderson K. Understanding place meanings for wilderness: Personal and community values at risk. International Journal of Wilderness. 2006;**12**:27-31

[7] Nelson KC, Monroe MC, Johnson JF, Bowers A. Living with fire: Homeowner assessment of landscape values and defensible space in Minnesota and Florida, USA. International Journal of Wildland Fire. 2004;**13**:413-425

[8] Eriksen C, Prior T. The art of learning: Wildfire, amenity migration and local environmental knowledge. International Journal of Wildland Fire. 2011;**20**:612-624

[9] United States Department of Agriculture and the U.S. Department of the Interior. National Fire Plan: A Report to the President in Response to the Wildfires of 2000. Washington, DC: USDA & USDI; 2001

[10] United States Department of Agriculture and the U.S. Department of the Interior. (2009) Quadrennial Fire Review 2009: Final Report. www.iafc.org/files/wild_QFR2009Report.pdf [Accessed: December 4, 2013]

[11] Bright AD, Newman P. How forest context influences the acceptability of prescribed burning and mechanical thinning. In: McCaffrey SM, editor. The Public and Wildland Fire Management: Social Science Findings for Managers. USDA Forest Service, Northern Research Station Gen. Tech. Rep. NRS-1. Newtown Square, PA; 2006. pp. 47-52

[12] Brunson MW, Shindler BA. Geographic variation in social acceptability of wildland fuels management in the western United States. Society & Natural Resources. 2004;**17**:661-678

[13] McCaffrey S, Kumagai Y. No need to reinvent the wheel: Applying existing social science theories to wildfire. In: Daniel TC, Carroll MS, Moseley C, Raish C, editors. People, Fire, and Forests: A Synthesis of Wildfire Social Science. Corvallis, OR: Oregon State University Press; 2007. pp. 12-36

[14] Monroe MC, Long AJ, Marynowski S. Wildland fire in the southeast: Negotiating guidelines for defensible space. Journal of Forestry. 2003;**101**:14-19

[15] Ryan RL, Wamsley MB. Public perceptions of wildfire risk and forest management in the central pine barrens of Long Island (USA). The Australasian

Journal of Disaster and Trauma Studies. 2008;**2008**(2):34-48

[16] Weible C, Sabatier P, Nechodom M. No sparks fly: Policy participants agree on thinning trees in the Lake Tahoe Basin. Journal of Forestry. 2005;**103**:5-9

[17] McCaffrey S, Toman E, Stidham M, Shindler B. Social science research findings related to wildfire management: An overview of recent findings and future research needs. International Journal of Wildland Fire. 2013;**22**:15-24

[18] Manfredo MJ, Fishbein M, Haas GE, Watson AE. Attitudes toward prescribed fire policies. Journal of Forestry. 1990;**88**:19-23

[19] Vining J, Merrick MS. The influence of proximity to a National Forest on emotions and fire-management decisions. Environmental Management. 2008;**41**:155-167

[20] Blanchard B, Ryan RL. Managing the wildland urban interface in the northeast: Perceptions of fire risk and hazard. Northern Journal of Applied Forestry. 2007;**24**:203-208

[21] Carroll MS, Cohn PJ, Seeholtz DN, Higgins LL. Fire as a galvanizing and fragmenting influence on communities: The case of the Rodeo-Chediski fire. Society and Natural Resources. 2005;**18**:301-320

[22] McCaffrey S. Thinking of wildfire as a natural hazard. Society & Natural Resources. 2004;**17**:509-516

[23] Steelman TA. Addressing the mitigation paradox at the community level. In: Martin W, Raish C, Kent B, editors. Wildfire Risk: Human Perceptions and Management Implications. Washington, DC: Resources for the Future; 2008. pp. 64-80

[24] Luloff AE, Field DR, Krannich RS, Flint CG. A matrix approach for understanding people, fire, and forests. In: Daniel TC, Moseley CM, Raish C, editors. People, Fire and Forests. Corvallis, OR: Oregon State University Press; 2007. pp. 207-216

[25] Oliviera P, Panagopoulos T. Post-fire landscape rehabilitation at the protected area of Montejunto, Portugal. In: 1st WSEAS International Conference on Landscape Architecture; June 11-13; Algarve, Portugal; 2008. Available from: https://www.researchgate.net/publication/268428599

[26] Daniel TC, Weidemann E, Hines D. Assessing public tradeoffs between fire hazard and scenic beauty in the wildland/urban interface. In: Jakes PJ, editor. Homeowners, Communities, and Wildfire: Science Findings from the National Fire Plan. USDA Forest Service, North Central Research Station Gen. Tech. Rep. NRS-1. St. Paul, MN; 2003. pp. 36-44

[27] Jakes PJ, Nelson KC. Community interactions with large wildfire events: Critical initiatives prior to the fire. In: Daniel TC, Carroll MS, Moseley C, Raish C, editors. People, Fire, and Forests: A Synthesis of Wildfire Social Science. Corvallis, OR: Oregon State University Press; 2007. pp. 91-103

[28] Loures L, Dias S, Ramos T, Nunes J, Viegas A. Severe forest fires: Assessment methods and reclamation techniques. International Journal of Energy and Environment. 2012;**6**(4):424-432

[29] Winter G, Fried JS. Homeowner perspectives on fire hazard, responsibility, and management strategies at the wildland-urban interface. Society & Natural Resources. 2000;**13**:33-49

[30] Moreira F, Vaz P, Catry F, Silva JS. Regional variations in wildfire susceptibility of land-cover types in Portugal: Implications for landscape management to minimize fire hazard.

International Journal of Wildland Fire. 2009;**18**:563-574

[31] Radeloff VC, Hammer RB, Stewart SI, Fried JS, Holcomb SS, McKeefry JF. The wildland-urban interface in the United States. Ecological Applications. 2005;**15**:799-805

[32] Absher JD, Vaske JJ, Bright AD. Basic beliefs, attitudes, and social norms regarding wildland fire management in Southern California. In: Chavez DJ, Absher JD, Winter PL, editors. Fire Social Science Research from the Pacific Southwest Research Station: Studies Supported by National Fire Plan Funds. USDA Forest Service, Pacific Southwest Research Station Gen. Tech. Rep. PSW-209. 2008. pp. 45-56

[33] Monroe MC, Pennisi L, McCaffrey S, Mileti D. Social Science to Improve Fuels Management: A Synthesis of Research Relevant to Communicating with Homeowners about Fuels Management. USDA Forest Service, North Central Research Station Gen. Tech. Rep. GTR-NC-257. St. Paul, MN; 2006

[34] Jacobson SK, Monroe MC, Marynowski S. Fire at the wildland interface: The influence of experience and mass media on public knowledge, attitudes, and behavioral intentions. Wildlife Society Bulletin. 2001;**29**:929-937

[35] Toman E, Shindler B. Wildland fire and fuel management: Principles for effective communication. In: McCaffrey SM, editor. The Public and Wildland Fire Management: Social Science Findings for Managers. USDA Forest Service, Northern Research Station Gen. Tech. Rep. NRS-1. Newtown Square, PA; 2006. pp. 111-123

[36] Knotek K, Watson A, Borrie WT, Whitmore JG, Turner D. Recreation visitor attitudes towards management-ignited prescribed fires in the Bob Marshall Wilderness Complex, Montana. Journal of Leisure Research. 2008;**40**:608-618

[37] McFarlane BL. Human dimensions of fire management in the wildland-urban interface: A literature review. In: Hirsch KG, Fuglem P, editors. Canadian Wildland Fire Strategy: Background Syntheses, Analyses, and Perspectives. Edmonton, AB: Canadian Council of Forest Ministers; 2006. pp. 27-36

[38] Gordon JS, Stedman RC, Matarrita-Cascante D, Luloff AE. Wildfire perception in rapid growth communities. Rural Sociology. 2010;**75**:455-477

[39] Moreira F, Viedma O, Arianoutsou M, Curt T, Koutsias N, Rigolot E, et al. Landscape – Wildfire interactions in southern Europe: Implications for landscape management. Journal of Environmental Management. 2011;**92**:2389-2402

[40] Bowker JM, Lim SH, Cordell HK, Green GT, Rideout-Hanzak S, Johnson CY. Wildland fire, risk, and recovery: Results of a national survey with regional and racial perspectives. Journal of Forestry. 2008;**106**:268-276

[41] Shindler BA, Toman E, McCaffrey SM. Public perspectives of fire, fuels, and the Forest Service in the Great Lakes region: A survey of citizens in Minnesota, Wisconsin, and Michigan. International Journal of Wildland Fire. 2009;**18**:157-164

[42] Smith M, Krannich R. Culture clash revisited: Newcomer and longer-term residents' attitudes toward land use, development, and environmental issues in rural communities in the Rocky Mountain West. Rural Sociology. 2000;**65**:396-421

[43] McCaffrey SM, Olsen CS. Research Perspectives on the Public and Fire Management: A Synthesis of Current Social Science on Eight Essential

Questions. USDA Forest Service, Northern Research Station Gen. Tech. Rep. NRS-104. Newtown Square, PA; 2012

[44] Beck U. World risk society. Cambridge: Polity; 1999

[45] Paveglio TB, Carroll MS, Absher J, Robinson E. Symbolic meaning of wildland fire: A study of residents in the US Inland Northwest. Society and Natural Resources. 2011;**24**:18-33

[46] Greider T, Krannich RS, Berry EH. Local identity, solidarity, and trust in changing rural communities. Sociological Focus. 1991;**24**:263-282

[47] Drabek TE, Key WH, Erickson PE. The impact of disaster on kin relationships. Journal of Marriage and Family. 1975;**37**:481-494

[48] Tierney KJ, Lindell MK, Perry RW. Facing the Unexpected: Disaster Preparedness and Response in the United States. Washington, DC: Joseph Henry Press; 2001

[49] Brennan MA, Luloff AE, Finley JC. Building sustainable communities in forested regions. Society and Natural Resources. 2005;**18**:779-789

[50] Silva JS, Vaz P, Moreira F, Catry F, Rego FC. Wildfires as a major driver of landscape dynamics in three fire-prone areas of Portugal. Landscape and Urban Planning. 2011;**101**:349-358

[51] Krannich RS, Humphrey CR. Using key informants in comparative community research. Sociological Methods and Research. 1986;**14**:473-493

[52] Heckathorn DD. Respondent-driven sampling II: Deriving valid population estimates from chain-referral samples of hidden populations. Social Problems. 2002;**49**:11-34

[53] Elmendorf WF, Luloff AE. Using qualitative data collection methods when planning for community forests. Journal of Arboriculture. 2001;**27**:139-151

[54] Creswell JW. Qualitative Inquiry and Research Design: Choosing among Five Traditions. Thousand Oaks, CA: Sage; 1998

[55] Toman E, Stidham M, Shindler B, McCaffrey S. Reducing fuels in the wildland urban interface: Community perceptions of agency fuels treatments. International Journal of Wildland Fire. 2011;**20**:340-349

[56] Brenkert-Smith H. Homeowners' perspectives on the parcel approach to wildland fire mitigation: The role of community context in two Colorado communities. Journal of Forestry. 2011;**109**:193-200

[57] McCaffrey S. Understanding public perspectives of wildfire risk. In: Martin W, Raish C, Kent B, editors. Wildfire Risk: Human Perceptions and Management Implications. Washington, DC: Resources for the Future; 2008. pp. 11-22

[58] Vaske JJ, Absher JD, Bright AD. Salient value similarity: Social trust and attitudes toward wildland fire management strategies. Human Ecology Review. 2007;**14**:223-233

[59] Sturtevant V, Jakes P. Collaborative planning to reduce risk. In: Daniel TC, Moseley CM, Raish C, editors. People, Fire and Forests. Corvallis, OR: Oregon State University Press; 2007. pp. 44-63

[60] Calkin DE, Cohen JD, Finney MA, Thompson MP. How risk management can prevent future wildfire disasters in the wildland-urban interface. PNAS. 2014;**111**(2):746-751

[61] Garkovich L. Community situational analysis. In: Foundations of Practice. Mississippi State, MS: Southern Rural Development Center. Retrieved from: http://srdc.msstate.edu/

fop/levelone/trainarc/09fall.html; 2009 [Accessed: November 2, 2014]

[62] Fleeger WE. Collaborating for success: Community wildfire protection planning in the Arizona White Mountains. Journal of Forestry. 2008;**106**:78-82

[63] Jakes P, Kruger L, Monroe M, Nelson K, Sturtevant V. Improving wildfire preparedness: Lessons from communities across the US. Human Ecology Review. 2007;**14**:188-197

[64] Blatner KA, Carroll MS, Daniels SE, Walker GB. Evaluating the application of collaborative learning to the Wenatchee fire recovery planning effort. Environmental Impact Assessment Review. 2001;**21**:241-270

[65] Daniel TC, Daniel T, Carroll M, Moseley C, Raish C, editors. People, Fire and Forests. Corvallis, OR: Oregon State University Press; 2007

Chapter 6

# Coastal Landscapes of Peninsular Malaysia: The Changes and Implications for Their Resilience and Ecosystem Services

*Jamilah Mohd Salim, Maizatul Azwa Radzi, Sheriza Mohd Razali and Fadzilah Majid Cooke*

## Abstract

Coastal landscapes are not only supporting the most productive and ecologically valuable ecosystem but are also fast changing, caused by both anthropogenic and natural processes. Changes in the form of diminishing vegetation cover, water body and increasing urbanization in Terengganu, East Coast of Peninsular Malaysia, for the years of 2000 and 2017 were assessed using Moderate Resolution Imaging Spectroradiometer satellite (MODIS) product. Images were processed based on Erdas Imagine software and then projected to World Geodetic System (WGS 84) coordinates based on ArcGIS 10.0. Significant reduction is detected in vegetation cover, from 46.5% in the year 2000 to 26.6% in 2017, coinciding with an increase in urban areas (from 3.3 to 33.6%). Changes due to urbanization raise concern over the loss of coastal landscape and may impact its resilience, so it may no longer be able to provide key ecosystem services. This understudied ecosystem deserves to be conserved for its ecosystem services. The paper argues that looking at the data presented, the resilience or the capacity of the Terengganu coastal landscape in maintaining its ecosystem services in the near future might have been compromised. Recommendations on how these valuable landscapes could be best conserved for social and ecological sustainability are put forward.

**Keywords:** coastal ecosystem, coastal wetland, dune landscape, ecosystem service, land use change, resilience, socioecological landscape

## 1. Introduction

Sustainable development is a key agenda for present and future development and perhaps global concerted effort to reduce.

The concept of resilience in relation to ecology and ecosystem is defined as the ability of the ecosystem to absorb the disturbance without shifting to an alternative state and losing function and services [1]. It is often used to describe the characteristic features of a system that are related to sustainability, and the concept of resilience is used in various interdisciplinary works, particularly in addressing the interactions between people and nature. Resilience is also associated with the capacity of the ecosystem to undergo disturbance and maintain its functions and controls [2]. For example, changes in grass species in the rangeland reduce its capacity to continue functioning ecologically (such as in water use and nutrient cycle) under heavy grazing by animals [3]. Ecosystem resilience can play a prime role in maintaining conditions that will sustain the provision of ecosystem services that contribute to the human well-being, in this case the well-being of coastal communities. The resilience of the ecosystem could directly affect the socioecological system. The objective of this chapter is to discuss the resilience of the coastal ecosystem of Terengganu, East Coast of Peninsular Malaysia, based on land use cover changes in Terengganu between years 2000 and 2017. Threats faced by the coastal ecosystem of Terengganu that may affect system resilience and ecosystem services are also discussed.

The ecosystem services concept was coined to address various benefits and values provided to humankind by ecosystems, which arise from ecological functions and biodiversity [4, 5]. The benefits and values could be direct or indirect, covering a wide range of vital goods and services that are classified into various ecological functions, for example, the provisioning service (such as providing goods or raw materials), regulatory services (such as air and water purification, water and nutrient cycling and regulation, soil formation and retention, atmospheric carbon sequestration) and supporting services. The last is the prerequisite for three other sets of ecosystem services (provisioning, regulating and cultural). However, the classification and typology of ecosystem services are varied and sometimes debatable in terms of application and relevance [6]. Nevertheless, ecosystem services as a concept are acknowledged to be an important tool to raise awareness on ecosystem's importance, particularly through identification of the goods and services made available by the ecosystem. The quantification of ecosystem services provide a monetary dimension, creating a potential link between biodiversity conservation and market value. In this chapter, we identify and discuss key ecosystem services of the coastal ecosystem of Terengganu and how it might have been interrupted by the changes.

Worldwide, coastal landscapes change tremendously due to urbanization and various other pressures both from anthropogenic-based and natural processes. Coastal landscapes are among the most densely populated zone as this zone provides enormous values and services to human population. Coastal ecosystem is commonly addressed together as 'estuarine and coastal ecosystems' (ECEs) due to their close connectivity and complexity in providing ecological services [7]. It not only holds high key economic values and vital ecosystem services but also supports integrated systems of social and ecological landscapes (SEL) [8]. At the global scale, coastal vegetation varies across geographical regions. In Malaysia and other tropical countries, there are three common vegetation types easily found in coastal zones, namely, mangroves, peat swamp forest and freshwater swamp. Coastal vegetation plays a significant role in stabilizing coastal ecosystems, for example, by modifying and stabilizing the physical environment [9]. The loss of coastal vegetation or changes in land use cover of terrestrial ecosystem could change the biomass and productivity leading to the changes in carbon cycling processes [10]. Coastal wetland reclamation causes loss in ecosystem services, for example, in Lianyungang Province in China [11]. Coastal wetland ecosystem varies in subtypes which comprises of estuaries, marshes, salt ponds, lagoons, mangroves, intertidal habitats and other coastal system subtypes. All wetland ecosystems vary in terms of unit value and ecosystem services rendered and even within the same eco-subtype; the unit value may vary with different space and time [4]. Thus, for the unit value of ecosystem services, different coastal wetland should be conserved and managed differently.

The East Coast of Peninsular Malaysia coastal plain is originated from marine-based deposit arranged in a series of ridge and depression parallel to the shoreline [12]. This soil formation is classified as "beach ridges interspersed with swales"

**(BRIS)** or locally known as *tanah beris* or *tanah bris*, where *tanah* is soil in Malay. BRIS soil formation is an oligotrophic type of soil, infertile and unsuitable for agriculture [13]. BRIS soil composes of more than 90% of sand (**Figure 1**). Despite that, it supports adapted and distinct vegetation formation which differs from common lowland tropical rainforest [14]. BRIS soil system occurs more abundant in the state of Terengganu relative to other states in the East Coast of Peninsular Malaysia, namely, Pahang, Kelantan and Johor. Coincidently, Terengganu also poses the longest coast compared to the other east coast states. Being one of the main oil-producing states in Malaysia, the coastal plain of Terengganu is already well developed with coastal road, settlement and infrastructures. However, it is a worrying fact that about 30% of Malaysia's coastlines are exposed to erosion [15]. More worrying is the fact that coastline erosion or accretion is not only caused by large monsoon waves but also by a more complicated interaction of offshore bottom bathymetry and island shelters, whereby these two components become a site-specific factor that helps to focus or disperse the energy of the monsoon waves to localize erosion or accretion [16]. Coastal erosion further became more frequent, subsequent to major sea reclamation for an airport runway upgrading in 2008 [17].

In the past decades, Terengganu has rapidly developed its overall economy through the federal government's East Coast Economic Region (ECER) Master Plan that was launched in 2008 headed by the East Coast Economic Region Development Council (ECERDC) [18]. The development programs and projects, among others, aim to raise the income levels and reduce poverty of the Terengganu population by expanding employment prospects in the east coast regions. Many of the projects take place along the coast itself, for example, development of a new central business district (CBD) at the north and south estuary of the Kuala Terengganu town centre and the planned development of the 600-km east coast rail line (ECRL) planned for linking key industrial hubs in Terengganu with Kuantan Port in Pahang and Port Klang in Selangor, both to its south. Some developments in Terengganu are located on the shoreline itself, for example, hipster concept restaurants along the coast of Tok Jembal, in Kuala Nerus district. Looking at this trend, the future outlook for Terengganu coastal ecosystem is rather challenging based on the worrying fact that about 30% of Malaysia coastlines are exposed to erosion [15]. Terengganu coastline erosion or accretion is not only caused by large monsoon wave but also by a more complicated interaction of offshore bottom bathymetry and island shelters [19]. Coastal erosion then becomes more frequent as a result of major sea reclamation for an airport runway upgrading in 2008 [17]. Further development in the coastal

**Figure 1.**
*Examples of typical soil series (Rudua and Rhu Tapai soil series) under beach ridges interspersed with swales (BRIS) system in the East Coast of Peninsular Malaysia compose more than 90% of sand.*

zone of Terengganu needs in-depth analysis on the current physical setting to reduce impact on coastal environment and community. This paper discusses coastal changes in Terengganu by looking at land use changes in terms of vegetation cover, urbanization and water body from the years 2000 to 2017 and the impact of these changes to Terengganu coastal ecosystem resilience and ecosystem services.

## 2. Changes in coastal landscapes and implications for ecosystem services

### 2.1 Ecosystem services of coastal landscapes

Among the most significant ecosystem services of coastal landscapes is perhaps coastal protection. The coastlines of eastern Peninsular Malaysia are directly exposed to the South China Sea's strong winds and dynamic coastal processes. Coastal vegetation acts as a first line of defence from physical elements of wind and wave due to exposure to the annually occurring northeast monsoon. At the same time, coastal vegetation holds together structurally loose coastal sandy soil. The Terengganu coast is also blessed with a prominent stretch of pure stand of *Melaleuca cajuputi* trees which barricade strong wind, protecting its coast and inland [20]. Having soil attributes of beach ridge system or BRIS, many parts of the coast of Terengganu also support a seasonal freshwater swamp or often addressed locally as *paya gelam* (in Malay) or gelam swamp as this swamp is dominated by gelam or *M. cajuputi*. This swamp is a seasonal wetland where its volume of water is contributed mainly by rain and to some extent by the overflow of small river tributaries during the monsoon season. Gelam swamp could support up to a 2–4 metre depth of water which is closely related to its function of mitigating flood in coastal areas and inland, particularly in the rainy season during monsoon months. Swale element in beach ridge soil of the East Coast of Peninsular Malaysia coastal plain acts as a sponge to keep subterranean water source, thus regulating local hydrological cycle [21]. Supporting one of the rarest type of wetland, a freshwater seasonal wetland (e.g. in Tasik Berombak of Setiu Coast and Jambu Bongkok, Dungun) [22]. The BRIS soil formation system plays a critical role in the local hydrological cycle, since it stores underground water and a deep layer of sand (~15 m below ground as recorded in Tasik Berombak, Setiu, Terengganu) which then act as a natural water filter and storage for clean freshwater—an important source for nearby areas becoming a part of a complex hydrological system of the coastal plain [21]. Although this kind of regulatory services carried out by BRIS soil ecosystem is hardly visible, the effect on social resilience on the local community is profound. It plays a critical role in providing adequate amount of good quality freshwater to support local economic activities of the coastal community, for example, in the district of Setiu, where the brackish lagoon is heavily used for aquaculture activities.

Other than the hydrological aspect, some part of BRIS soil ecosystem is comprised of newly developed peat, which is an important form of carbon storage [23]. Soil carbon together with above and below ground biomass of plants is a very important carbon sink. Even though above ground carbon in the biomass of *M. cajuputi* on dune landscape of Terengganu is much lower than other common Malaysian tropical lowland forests [24], *M. cajuputi* tree stand still serves as an important local carbon stock that could help in mitigating climate change effect. Carbon fixed in the above and below ground biomass of *M. cajuputi* could help reduce carbon being released to the atmosphere, thus reducing the effect of global warming. The benefit of conserving forest for carbon stock is well discussed as part of many ecosystem services of forest [4]. Sparse natural vegetation growing on the coastal plain of Terengganu plays a vital role in stabilizing the loose structure of coastal soil, growing on both ridge (dry area) and depression (swales or waterlogged areas) of sand dunes. On the ridge, vegetation is growing in the clump to optimize soil resources needed for growth and development. Removal of natural vegetation either by natural (e.g. wild fire) or anthropogenic activities (e.g. legal and illegal sand mining) may cause coastal erosion, leaving the soil prone to be invaded by exotic invasive species of *Acacia mangium* (Fabaceae) or indigenous species *Catunaregam tomentosa* (Rubiaceae) [25]. *Acacia mangium* is not yet declared as invasive species in Malaysia, but its ability to negatively affect and alter nearby plant composition in its presence, particularly through its allelopathic effect, is well known [26]. *Acacia mangium* can easily invade BRIS ecosystem due to open canopy and low stature of its vegetation that grows in clumping pattern. The abundance of *A. mangium* mother trees in and around the coastal ecosystem of Terengganu facilitates the dispersion of this species. The seed of *A. mangium* is dispersed by birds and wind and easily germinates underneath vegetation clump. Many degraded BRIS soil ecosystems along the coast of Terengganu are already invaded and totally taken over by this species [4, 20, 25]. It is well acknowledged that invasive plant species can decrease resilience by reducing the biodiversity in the ecosystem that is being invaded and eventually will interrupt key ecosystem services provided by one ecosystem [27]. However, for the coastal landscape of Terengganu, the lack of interest and awareness from local authorities may have contributed to the lack of research funding to address this issue.

Provisioning services of the coastal landscape of the East Coast of Peninsular Malaysia are closely related to support livelihoods of its fishery communities, for example, the utilization of the most abundant plant resources, *M. cajuputi* (Gelam) wood and other parts. Woods of *M. cajuputi* are processed for charcoal and poles which are used as construction material and in scaffolding for small-scale construction such as for fishing jetty and port. The bark of gelam is traditionally used to seal boat walls (caulking) [28], assisted by waterproof properties of the bark. Gelam tree is also widely planted as ornamental tree in urban areas and public parks throughout the country. The potential value of gelam in provisioning service includes the use of gelam in greening effort [29]. The tiny and abundant seed can germinate and grow well into seedlings, or vegetatively it can propagate easily using its root suckers [30]. Fire resistance of this tree provides an advantage for using this species in restoration effort. In the wild and on BRIS ecosystem disturbed habitat, postfire recovery of gelam is quickly taking place by regenerating coppice shoots, which originated from its apical buds underneath the bark [31]. Gelam provides a renewable resource of woods and poles and potentially can be used to produce *cajuput* oil, a secondary compound from its leaves which may be useful for pharmaceutical industry. The 'cajuput' oil industry is surviving well in Indonesia [32] and Thailand [33]. However, similar industry is still untapped in Malaysia or Terengganu, possibly due to low essential oil content in its leaves, about <1% of it dry weight [34]. Although *M. cajuputi* has low yield of essential oil, it is still a promising natural plant extract and is a far more environmentally friendly consumer product to replace chemical-based products [35].

Indirect use of pure Gelam stand supports healthy populations of bees and stingless bees, giving a source of sought after honey, collected by the local fishermen as their side income [36]. In swampy part of coastal plain, gelam trees act as a key species in the swampy part of coastal plain, supporting a healthy population of freshwater fishes that are commonly caught by the locals for their ornamental (e.g. tigerbarb) and also for nutritional values (e.g. catfish, snake head and climbing perch). The fishes are abundant during the monsoon season in Terengganu. There are more than 60 species of ornamental freshwater fishes recorded in the riverine system and swamps of Terengganu [37, 38]. Other than supporting freshwater fishes, gelam swamp provides habitat for hydrophytes (submerged, emergent, floating rooted) and woody and nonwoody associated plants. Carnivorous plants of *Nepenthes*, *Drosera* and *Utricularia* are also common at the fringe of the swamp offering a view of a montane or heath kind of flora on the lowland that is easily accessible for ecotourism or showcase [39]. A far more puzzling flora in the gelam swamp of Terengganu is the occurrence of an endemic sedge species of Peninsular Malaysia, *Websteria confervoides* (Cyperaceae), which is so far only recorded in Lake Bera (Pahang) dan Jambu Bongkok, Dungun (Terengganu). This plant depends greatly on the existence of the coastal wetland of gelam swamp and only abundant during high water level (0.5–2 m) [24]. The mechanism of how this plant could maintain its population in the dry swamp after a long drought in the dry season or non-monsoon months is still understudied and worth exploring. In Malaysia in general, intensive research on forest and vegetation are primarily focused on the dipterocarp forest for the inland forest and mangroves in the coast. It is worrying that lack of research in this similar kind of vegetation on the coastal plain of Malaysia will contribute to the poor understanding on how this ecosystem function provides key ecosystem services. Consequently, lack of knowledge about the ecosystem function may prevent us from building the resilience of this disappearing coastal ecosystem.

The ridge areas on the dune which are dryer due to its loose sandy structure surprisingly support quite a number of adapted coastal vegetation [25], including more than 30 species of wild orchids [40] (**Figure 2**). Thus, the Terengganu coastal plain could be an important gene bank for wild orchids that could support commercial orchid industry, one of the option values under the total economic valuation (TEV) [5]. The aesthetic value of this coastal ecosystem together with its natural flora, fauna and landscapes could potentially be conserved and highlighted as one of the many ecotourism products for Terengganu to add to the economic benefit to the coastal communities. This value could be a monetary trade-off for conserving Terengganu BRIS ecosystem. With all the outlined ecological values, services and potentials, gelam forest is no doubt a valuable premise for Terengganu's coastal ecosystem resilience. Maintaining healthy Gelam forests will help maintain their ecological services for the benefit of the coastal environment that supports the livelihoods of coastal communities. Rather than being seen as unproductive and unimportant, gelam forest should be conserved for their values and services. Awareness on the importance of gelam forests to the sustainability of coastal ecosystem and people should be intensified. Factors contributing to the risk faced by the Gelam forest are outlined in the next section.

**Figure 2.**
*Natural vegetation on dry part (ridge) of BRIS soil ecosystem on Terengganu coast with a clumping pattern of vegetation (left image) and wild orchid species,* Phalaenopsis pulcherrima, *thriving well underneath vegetation clump (right image).*

## 2.2 Threats to gelam forest and coastal landscape of Terengganu

The coastal ecosystem of Terengganu is at risk of disappearing if there is no effort in conserving or managing this ecosystem in a sustainable way. Fragmentations of Terengganu coastal ecosystem are mainly due to reclaimation for housing or settlement on a private land, or a development of new township and infrastructure on the state owned land. This is primarily due to its strategic location along the main coastal road, as well as on the lower terrain. Failure in seeing the values of natural ecosystem, shadowed by the lack of value for agriculture, and BRIS soil ecosystem is considered as a barren land and wasteland that deserve to be converted to other land uses. This ecosystem is also threatened by illegal chemical and solid waste dumping, as observed in many areas along the coast of Setiu (north of Kuala Terengganu) and Marang (south of Kuala Terengganu) (**Figure 3**). The lack of public knowledge about the values of BRIS soil coastal ecosystem and low civic mindedness are identified as primary causes to this problem. Lack of human presence and visible activities in the ecosystem itself also encourage the act of illegal dumping. Frequent monitoring by local authority could help reduce the incidence of illegal waste dumping [20, 25].

BRIS soil vegetation can easily catch fire, particularly in non-monsoon months or drought season (**Figure 3**) which can be of natural process and human induced. High incidence of sunray and high temperature of sandy soil surface may initiate fire naturally. Fire can also occur simply from human reckless behaviour, for example, by throwing cigarette butts into the dry and sparse vegetation on BRIS soil ecosystem. There was an extensive fire occurrence recorded along Terengganu coast [41] and several places along coastal road in Setiu experiencing fire in 2016, coinciding with low rainfall and drought in 2014–2016 [30]. Fire is one of driven factors for ecological succession [42] and sometimes needed for vegetation regeneration [43]. However, with the presence of fire-adapted species, ecosystem resilience is negatively affected [44]. This brings us to the next threat faced by Terengganu

**Figure 3.**
*Threats to coastal ecosystem of Terengganu, frequent fire occurrence particularly during drought or non-monsoon months (top row images), illegal sand mining (bottom row, left image) and illegal dumping (bottom row, right image).*

coastal ecosystem, which is colonization of *A. mangium*. It is well noted that a slight modification to BRIS soil ecosystem on Terengganu coast leads to colonization of invasive *A. mangium* [20]. Many sites of BRIS soil vegetation has been replaced totally by *A. mangium* with no sign of natural vegetation underneath. On the other hand, the natural vegetation could be replaced totally by the ferocious spiny shrub of *Randia tomentosa* (Rubiaceae). Changes in plant composition reduce the resilience of the ecosystem, whereby it is shifted towards less diverse in species [1]. This eventually affects many ecosystem services related to plant's roles, for example, in regulating soil nutrient cycle and supporting animal diversity.

The other major threat to BRIS soil ecosystem of Terengganu is sand mining, which commenced a few years back when there was a high demand for sand from the Terengganu coast as it contains high-quality silica. Sand is mined illegally and possibly being transported to the other states or countries to meet the demand. The illegal and small-scale sand mines operated by removing small patches of sand, usually not that far from the coast itself. However, there is one site being mined with the size as big as football field near Lembah Bidong in Setiu district (**Figure 3**). Experimental study at this degraded site indicates that regeneration of natural vegetation is low and occurs at a very slow rate. Thus, illegal small-scale sand mining could be interfering with key ecosystem services of the coastal ecosystem due to removal of sand and vegetation. In the case of legal and large-scale sand mining, currently Terengganu has two sites of sand mining, privately operated and declared as not detrimental to the coastal environment. However, it is doubtful that the impact of sand mining to the coastal ecosystem is low; rather, the extent of the impact is still uncertain and unknown, as the sand mining is a newly emerging economic activity in Terengganu. The hope is that this industry will be well regulated and monitored by the authority to minimize its impact on the coastal environment.

### 2.3 Changes in the coastal landscapes of Terengganu (Years 2000–2017)

Based on images of Terengganu vegetation cover for years 2000 and 2017, it is clear that the coastal area of Terengganu is changing due to urbanization (**Figure 4**). Urban area has increased from about 3.3% in the year of 2000 to 33.6% in 2017 (**Table 1**). Even though the outline data does not specifically indicate differences contributed by the reduction of coastal area, it is clear that there is an increase in urbanization areas along the coast of Terengganu in 2017. Major changes to the Terengganu coastline begin in 2008 when parts of the sea off Terengganu were reclaimed for an airport runway upgrading [17]. Such major reclamation not only caused erosion but also halted the natural accretion process by disturbing sediment transport along the coastline [45]. Consequently, episodic erosion occurred in the northern part of the Terengganu coastline, and the most recent erosion occurs in Kampung Mengabang Telipot, north of Kuala Terengganu state capital [46].

Erosion and accretion are natural processes and part of ecological coastal dynamic. However, severe erosion fundamentally indicates failure of managing coastal zone when longshore sediment transport is interrupted by engineering works such as construction of groynes and breakwaters along the northern Terengganu coast [47]; most possibly it is happening in recent breakwater establishment along the coast of Terengganu (**Figure 5**). Other possible causes of erosion are removal of natural vegetation that can dissipate the wave energy, reduction of sediment supply from engineering works in rivers such as dams and barrages, sand mining from river bed and unregulated or uncontrolled dredging and sand mining activities in near shore areas. All of these factors seem to be part of the contributing agents to Terengganu coastal erosion. It is a prime challenge for the authority of the state of Terengganu to find a creative engineering technique to solve this

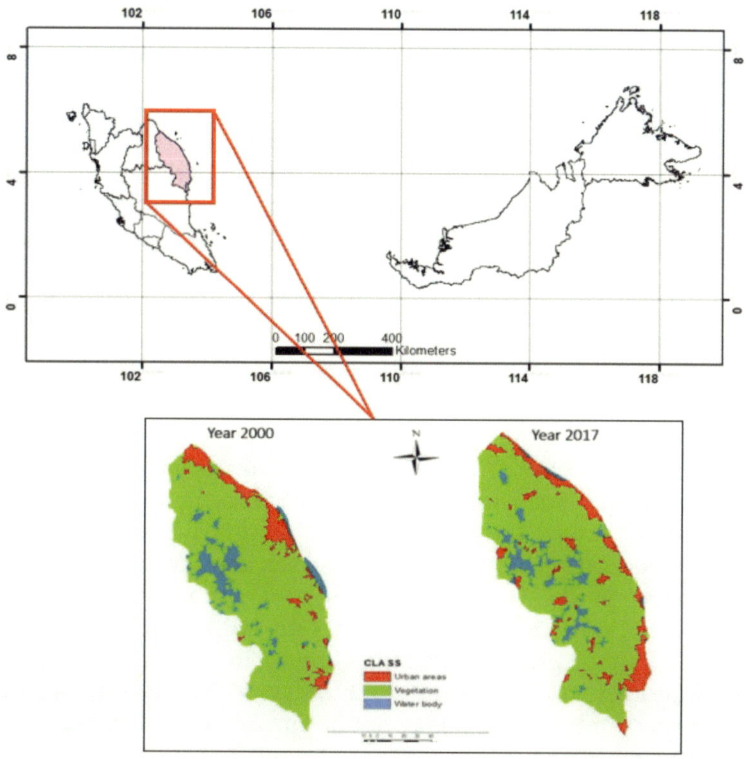


**Figure 4.**
*Map of Peninsular Malaysia (top row) and vegetation cover in the state of Terengganu, East Coast of Peninsular Malaysia, for the years 2000 and 2017 (bottom row). Image source: Land Process Distributed Active Archive Centre (LPDAAC).*

| Land Use Classification | Year 2000 | | Year 2006 | | Year 2017 | |
|---|---|---|---|---|---|---|
| | Hectare | Percentage | Hectare | Percentage | Hectare | Percentage |
| Water bodies | 15674.43 | 50.1 | 15477.64 | 53.2 | 13092.78 | 39.7 |
| Vegetation | 14547.42 | 46.5 | 15300.67 | 31.5 | 8787.50 | 26.6 |
| Urban area | 1038.49 | 3.3 | 4458.33 | 15.3 | 11094.25 | 33.6 |
| Total | 31260.34 | 100 | 29097.90 | 100 | 32974.53 | 100 |

*Note: Data in hectare are extracted from satellite images obtained from the Land Process Distributed Active Archive Centre (LPDAAC).*

**Table 1.**
*Vegetation cover for the state of Terengganu, Peninsular Malaysia, for years 2000, 2006 and 2017*

**Figure 5.**
*Coastal erosion along the Universiti Malaysia Terengganu (UMT) campus in Kuala Nerus district, north of Kuala Terengganu (left image), and breakwaters constructed to solve erosion along north of Terengganu (right images). White arrows mark extension of airport runway in 2008. Source: Media Kreatif UMT (left image) and Mr. Mokhtar Ishak (right image).*

complicated 'man versus nature' situation. To ensure the sustainability of the coast, significant efforts should be made to maintain ecological infrastructures or multifunctional network of ecosystem provided by coastal wetlands [11]. Considering the dynamics of the Terengganu coast, it is recommended that the coastal sustainable land use planning (SLUP) strategy be adopted. SLUP is evident to enhance coastal resilience, so that coastal ecosystem could continue to provide key ecosystem services, particularly for the benefit of the coastal community [8, 48, 49].

The reduction of vegetation cover in some parts of the coastal areas of Terengganu is possibly due to vegetation removal for aquaculture activities and settlement construction. In the coastal areas of Terengganu, apart from mangrove trees and associated plants, *M. cajuputi* (gelam) tree clearance is common. For example, in Kemaman and Setiu districts, pure stand gelam trees are cleared to make ways for township development and aquaculture complex, respectively. Vegetation clearance using heavy machineries is a common practice during land preparation for the construction of residential or commercial buildings. Should the sites happen to be on swampy or wet areas, sand or top soil is used to reclaim them before construction commence. In most of the state in Malaysia, regreening or revegetation of the developed areas is voluntary and not regulated. This could contribute to the loss of vegetation in newly developed urban areas. However, the reverse may happen whereby land is cleared for oil palm plantation, which then contributes to the increase in vegetation cover; albeit, oil palm plantations are a monocrop and not biodiverse. Therefore, oil palm plantations and natural stands of gelam may not be similar in quality and quantity of providing ecosystem services.

### 2.4 Impact of coastal landscape changes on ecosystem resilience and social environment

An interesting shift that has taken place in resilience thinking that is of relevance to this paper. The premise in resilient thinking that ecological resilience is key to the management of changes occurring in complex and dynamic systems of people and nature cannot be understood if there is little understanding of the social drivers of change that contributes to that ecological resilience [50]. 'People do change the resilience of ecological systems' ([50]:p.428).

Complexity and diversity as well as fragility are deemed to be the characteristics of both social and natural systems so that responses to interventions or encroachments are unpredictable. Ecological resilience taken to mean the capacity for renewal in a dynamic environment is required in order for the system to respond to the social drivers of change, albeit in an unpredictable manner. The major social drivers of change that are most mentioned in the literature, because of their generalized presence in landscapes and regions around the world, are acknowledged to be unsustainable land use, abandonment and urbanization [51]. These some drivers are also occurring in the coastal landscapes of Terengganu, as mentioned in earlier sections of this paper.

The tendency to focus on man-made degradation of ecosystems in studies of resilience has been criticized. Instead, it is recommended that solutions should be focused on creative processes of accumulating natural capital developed and should include their intangible values. This is also due to the assessment practices that commonly focus on visible or tangible change (biodiversity loss, brittle stability, of 'an accident waiting to happen') [50]. Examples of intangible values are those associated with biodiversity conservation (for ecotourism, or for ecosystems services it renders to human populations).

Since human well-being is also linked to non-tangible (non-market values), there has been an increasing interest in cultural landscapes (heritage places, regions that have iconic value for identity formation—nationalism—such as Mount Kinabalu for Sabahans of Malaysia, the pastoral landscapes of England and many more). These non-market values are broadly captured by the literature on 'cultural values' [52, 53]. We will focus on one element of cultural values, namely, identity strengthening, which is linked to a sense of place. According to [52] the concept of a sense of place 'embeds all dimensions of peoples' perceptions and interpretations of the environment, such as attachment, identity or symbolic meaning, and has the potential to link social and ecological issues'. An example of a sense of place, in this instance the link between the Terengganu coastal system and the identity of fishers, can be taken as an example as below.

Livelihood security of artisanal fishers of Terengganu depends on the sea—near shore and further in the open South China Sea. However, the sea provides more than livelihoods to fishers. Anecdotal evidence from newspaper clippings indicate that despite risks from coastal erosion, many local residents find it difficult to leave because they claim that they have nowhere to go [54]. As well, among artisanal fishers of the Setiu wetlands in Terengganu, despite risks from weather disturbances and being employed in more stable occupations such as in aquaculture, many fishers maintain their fishing trips out to at least three to four times a week except during severe monsoons [55]. This maintenance of their connection with the sea is what distinguishes those who consider themselves as 'real' fishers versus those who are not (including those who have boats and equipment but do not pursue fishing seriously). The sea then carries the intangible value of providing some fishers with a mechanism for strengthening their cultural identity. Similar findings on the effects of place (whether marine or terrestrial) and identity are evident in many studies around the world [56]. For example, in Sabah, Bajau fishers identify themselves with the inland sea surrounding the Banggi Island chain and their identity found strength in seaweed cultivation, despite the fact that fishing as an activity provided them with a higher return for hours worked than labour intensive seaweed cultivation [57].

The bio-security of Gelam forests depends on the degree of its resilience as well as the social resilience of the local communities that have lived alongside them or who are benefitted from the health of these forests. The ecosystem services provided by the wetlands and the dry swamp of Gelam forests are including uses in the construction of sea-going fishing boats. Freshwater fishes found in flooded lakes and riverine systems during the annual monsoon season provide extra source of nutrition to local communities as outlined in earlier sections of this paper. But the reverse provision of services by local communities, through their local knowledge in the sustainable management (through use) of natural resources from inland forests and seas, has not been well researched.

Consequently, a lot more research needs to be done on how local communities form knowledge about their landscapes. Secondly, given the understanding that throughout history there are very few landscapes in the world that have not been shaped by local communities [54], to what extent has local knowledge shaped the characteristics of the gelam landscape? These are valid questions to ask because despite the transformation of landscapes by drivers of development as the Terengganu coast has been, certain cultural values are not totally lost as viewed in fishers' identity and place. As to why local knowledge research is important, there is a consensus that environmental degradation is not amenable for its solution to one body of knowledge alone but from a variety of knowledge types and disciplines.

## 3. Conclusion

There is a reduction in vegetation cover in Terengganu from the years 2000 and 2017, and it coincides with the increase in coverage of urban areas. Even though our data do not particularly reflect specific changes to coastal areas, this reduction in vegetation cover deserves to be addressed. It is time that the complexity of coastal ecosystem be valued as a social ecological landscape. Sustainable land use planning (SLUP) may be a good model to be adopted in managing coastal ecosystem of Terengganu. Sustainable solutions should be applied to aim for social, economic and environmental benefits. In-depth research on each component of social and ecological system and their connectivity should be enhanced to further understand coastal ecosystem resilience and assist the authority in the planning and managing of coastal ecosystem [58]. Better valuation of the landscape could be conducted to include general public perception analysis in the development planning [59]. Local knowledge of the ecosystem ought to be encouraged for their value to planning.

## Acknowledgements

The authors would like to acknowledge the Director of Institute of Tropical Biodiversity and Sustainable Development (Bio-D Tropika) of Universiti Malaysia Terengganu, Dato' Prof. Dr. Mohd Tajuddin Abdullah, for his support and encouragement in organizing the 20th Colloquium of Malaysia and Singapore Society of Australia (MASSA2018), February 8–9, 2018, where parts of this write up are presented. We would like to thank Media Kreatif UMT for the image of erosion along Terengganu coastline and Mr. Mokhtar Ishak for the drone images of breakwater along UMT beaches.

## Conflict of interest

There is no conflict of interest in this publication.

## Author details

Jamilah Mohd Salim[1,2,3]*, Maizatul Azwa Radzi[3], Sheriza Mohd Razali[2] and Fadzilah Majid Cooke[1]

1 Institute of Tropical Biodiversity & Sustainable Development (Bio-D Tropika), Universiti Malaysia Terengganu, Kuala Nerus, Terengganu, Malaysia

2 Institute of Tropical Forestry and Forest Products (INTROP), Universiti Putra Malaysia, Serdang, Selangor, Malaysia

3 School of Marine and Environmental Sciences, Universiti Malaysia Terengganu, Mengabang Telipot, Kuala Nerus, Terengganu, Malaysia

*Address all correspondence to: jamilah@umt.edu.my

## References

[1] Carpenter S, Walker B, Anderies JM, Abel N. From metaphor to measurement: Resilience of what to what? Ecosystems. 2001;**4**:765-781

[2] Holling CS. Resilience and stability of ecological systems. Annual Review of Ecology and Systematics. 1973, 1973;**4**:1-23

[3] Walker BH, Kinzig A, Langridge J. Plant attribute diversity, resilience, and ecosystem function: The nature and significance of dominant and minor species. Ecosystems. 1999;**2**:1-20

[4] Costanza R, de Groot R, Sutton P, van der Ploeg S, Anderson SJ, Kubiszewski I, et al. Changes in the global value of ecosystem services. Global Environmental Change. 2014;**26**:152-158

[5] Millennium Ecosystem Assessment (MEA). Ecosystems and Human Well-being: Synthesis Report. Washington DC: Island Press; 2005. 100p

[6] Pedrono M, Locatelli B, Ezzine-de-Blas D, Pesche D, Morand S, Binot A. Impact of climate change on ecosystem services. In: Torquebiau E, editor. Climate Change and Agriculture Worldwide. 2016. DOI: 10.1007/978-94-017-7462-8_19

[7] Barbier EB, Sally D, Hacker SD, Kennedy C, Koch EW, Stier AC, et al. The value of estuarine and coastal ecosystem services. Ecological Monographs. 2011;**201**:169-193

[8] Kim M, You S, Chon J, Lee J. Sustainable land-use planning to improve the coastal resilience of the social-ecological landscape. Sustainability. 2017;**9**:1086. DOI: 10.3390/su9071086

[9] Gracia A, Nelson Rangel-Buitrago G, Oakley JA, Williams AT. Use of ecosystems in coastal erosion management. Ocean and Coastal Management. 2018;**156**:277-289. DOI: 10.1016/j.ocecoaman.2017.07.009. ISSN 0964-5691

[10] IPCC. Land-use, Land-use change and forestry. In: Watson RT, Noble IR, Bolin B, et al., editors. A Special Report of the IPCC. Cambridge: Cambridge University Press; 2000

[11] Sun X, Li Y, Zhu X, Cao K, Feng L. Integrative assessment and management implications on ecosystem services loss of coastal wetlands due to reclamation. Journal of Cleaner Production. 2017;**163**:S101-S112. DOI: 10.1016/j.jclepro.2015.10.048 0959-6526

[12] Mohd Ekhwan HT, Mazlin BM, Muhammad Barzani G, Nor Azlina AA. Analysis of the physical characteristics of BRIS soil in coastal Kuala Kemaman, Terengganu. Research Journal of Earth Sciences. 2009;**1**:1-6

[13] Lim JS. National Report for the UNCCD Implementation: Combating Land Degradation and Promoting Sustainable Land Management in Malaysia. Kuala Lumpur: Department of Agriculture; 2002

[14] Jamilah MS, Nur-Faiezah AG, Siti Kehirah A, Siti Mariam MN, Razali MS. Woody plants on dune landscape of Terengganu, Peninsular Malaysia. Journal of Tropical Forest Science. 2014;**26**(20):267-274

[15] Ong JE. Vulnerability of Malaysia to Sea-Level Change [Internet]. 2001. Available from: http://citeseerx.ist.psu.edu/viewdoc/summary?doi=10.1.1.197.30339 [Accessed: May 1, 2016]

[16] Husain ML, Yaakob R, Saad S. Beach erosion variability during a northeast monsoon: The Kuala Setiu coastline,

Terengganu, Malaysia. Pertanika Journal of Science and Technology. 1995;**3**(2):337-348

[17] Muslim AM, Foody GM, Atkinson PM. Shoreline mapping from coarse–spatial resolution remote sensing imagery of Seberang Takir, Malaysia. Journal of Coastal Research. 2007;**23**:1399-1408. DOI: 10.2112/04-0421.1

[18] ECERDC [Internet]. 2018. Available from: http://www.ecerdc.com.my/en/master-plan/overview/ [Accessed: September 1, 2018]

[19] Husain ML. Open sandy beach morphology and morphodynamic as response to seasonal monsoon in Kuala Terengganu, Malaysia. Journal of Coastal Research: Special Issue 75. In: Proceedings of the 14th International Coastal Symposium, Sydney, 6-11; March 2016: pp. 1032-1036. DOI: 10.2112/SI75-207.1

[20] Jamilah MS, Norhayati Y, Kasawani I, Nursalwa B, Razali MS. Brief floristic descriptions of Jambu Bongkok Forest Reserve, BRIS ecosystem, Melaleuca swamp of Terengganu. In: Jamilah MS, Faridah M, editors. A Biological Assessment of Jambu Bongkok Forest Reserve. Terengganu and Nearby Ecosystem: Universiti Malaysia Terengganu; 2011. pp. 1-24

[21] Sathiamurthy E. Setiu River Basin: A brief account of the hydrologic connections of Setu Wetlands. In: Faridah M, Jamilah MS, Jarina MJ, Rohani S, editors. Species, Ecosystem and Livelihood. Universiti Malaysia Terengganu; 2015. pp. 101-107

[22] Salim JM, Mohamad F, Shahrudin R. Setiu: More than a Wetlands. In: *Setiu Wetland,* In *Species, Ecosystem and Livelihood*, Faridah Mohamad, Jamilah Mohd Salim, Jarina Mohd Jani & Rohani Shahrudin (eds.). Setiu Wetland: Species, Ecosystem and Livelihoods. Universiti Malaysia Terengganu. 2015. pp. 87-100

[23] Peter Abs Grootjans. Personal Communication in 2016. University of Groningen/Radboud University of Nijmegen, The Netherlands

[24] Jamilah MS. Four year climatic parameters influence on the abundance and survival of Websteria Confervoides (Submerged Cyperaceae), a Wetland Dependent Species. In: Paper Presented at the 10th Flora Malesia Symposium, Royal Botanic Garden Edinburgh, United Kingdom, 11-15 July 2016

[25] Jamilah MS, Nur-Faiezah AG, Siti Kehirah A, Siti Mariam MN, Razali MS. Woody plants on dune landscape of Terengganu, Peninsular Malaysia. Journal of Tropical Forest Science. **2014**:267-274

[26] Mustafa Kamal MS, Shamsul AB. Invasive plants in the Malaysian landscape. International Journal on Sustainable Tropical Design Research & Practice. 2006;**1**:41-48

[27] Eviner VT, Garbach K, Baty JH, Hoskinson SA. Measuring the effects of invasive plants on ecosystem services: Challenges and procpects. Invasive Plant Science and Management. 2012;**5**:125-136

[28] Burkill IH. A Dictionary of the Economic Products of the Malay Peninsula. Vols. 1 & 2. Governments of the Straits Settlements and Federated Malay States, London. Vol. I (A-H). pp. 1-1240 and Volume II (I-Z). pp. 1241-2444; 1935

[29] Erwin KL. Wetlands and global climate change: The role of wetland restoration in a changing world. Wetlands Ecology and Management. 2009;**17**:71-84. DOI: 10.1007/s11273-008-9119-1

[30] Jamilah MS, Rohani S, Abdul Shukor Y. Preliminary survey for fire

occurrence on the Bris ecosystem of Setiu, Terengganu during extended drought. In: Proceedingof Setiu Scientific Expedition Seminar, Universiti Malaysia Terengganu; 2016

[31] Aminuddin AH. The Effect of Fire and Mechanical Damages to *Melaleuca cajuputi* (L.) L. in Heath Vegetation of Terengganu [Undergraduate Thesis]. Bachelor of Applied Science in Biodiversity and Conservation Management, Faculty of Science and Technology, Kolej Universiti Sains dan Teknologi Malaysia; 2005

[32] Doran JC. *Melaleuca cajuputi* powell. In: Janson PCM, Westphal E, Wulijarni-Soetipto N, editors. Plant Resources of South East Asia: Essential-Oil Plants. Bogor, Indonesia: Procea Foundation; 1999. pp. 126-131

[33] Kim JH, Liu KH, Yoon Y, Sornnuwat Y, Kitirattrakarn T, Anantachoke C. Essential Leaf Oils from *Melaleuca cajuputi*. In: Palaniswamy UR, Craker LE, Gardner E, editors. Proc. WOCMAP III, Vol. 6. Traditional Medicine & Nutraceuticals. Acta Horticulture; 2005. p. 680

[34] Sainorudin MH, Rozaini MZH, Hamzah H, Saupi AAM, Norazemi NF, Ismail Z, et al. Preliminary study of sunscreen and anti-tyrosinase effect on microemulsion extract from melaleuca cajuputi essential oil using nonionic sur-factant. GSTF Journal of Chemical Sciences (JChem). 2015;**2**(1):1-9

[35] Bakar AA, Sulaiman S, Omar B, Ali RM. Evaluation of *Melaleuca cajuputi* powell (Family: Myrtaceae) extract in aerosol can against dengue vectors in the laboratory. Journal of Tropical Medicine and Parasitology. 2009;**32**:58-64

[36] Kassim M, Yusoff KM, Ong G, Sekaran S, Yusof MYBM, Mansor M. Gelam honey inhibits lipopolysaccha-ride-induced endotoxemia in rats through the induction of heme oxygenase-1 and the inhibition of cytokines, nitric oxide, and high-mobility group protein B1. Fitoterapia. 2012;**83**(6):1054-1059

[37] Amirrudin BA, Johari MN, Nurul Nadia R, Siti Aminah MS. Dragonfly of Hutan Simpan Jambu Bongkok, Terengganu, Peninsular Malaysia. 2011. In: Jamilah MS, Faridah M, editors. A Biological Assessment of Jambu Bongkok Forest Reserve. Terengganu and Nearby Ecosystem: Universiti Malaysia Terengganu; 2011. pp. 45-60

[38] Amirrudin BA, Muhammad Fahmi A, Johari MN. Freshwater fishes of Hutan Simpan Jambu Bongkok, Terengganu, Peninsular Malaysia. 2011. In: Jamilah MS, Faridah M, editors. A Biological Assessment of Jambu Bongkok Forest Reserve. Terengganu and Nearby Ecosystem: Universiti Malaysia Terengganu; 2011. pp. 31-43

[39] Jamilah MS, Nur-Atiqah MH, Nurul-Effadah S, Nur-Khairulhusna I. Pigmy *Drosera* (Droseraceae) from heath forest on BRIS soil at Setiu, Terengganu. Folia Malaysiana. 2009;**10**:33-40

[40] Siti Fatimah MI, Rusea G, Jamilah MS, Yien CYS. Orchids in coastal heath forest of Terengganu. In: Faridah M, Jamilah MS, Jarina MJ, Rohani S, editors. Species, Ecosystem and Livelihood. Universiti Malaysia Terengganu; 2015. pp. 51-56

[41] The New Straits Times, More Peat Swamp Forests were Destroyed by Fire in Terengganu, Aided by the Hot and Windy Weather Since End of Last Month [Internet]. 2014. Available from: http://www.pressreader.com/malaysia/the-borneo-post/20150310/282273843834258 [Accessed: September 9, 2018]

[42] Herath DN, Lamont BB, Enright NJ, Miller BP. Impact of fire on plant-species persistence in post-mine restored and natural shrubland communities in southwestern Australia. Biological Conservation. 2009;**142**:2175-2180

[43] Ellsworth LM, Wrobleski DW, Kauffman JB, Reis SA. Ecosystem resilience is evident 17 years after fire in Wyoming big sagebrush ecosystems. Ecosphere. 2016;**7**(12):1-12. DOI: 10.1002/ecs2.1618

[44] Mack MC, D'Antonio CM. Impacts of biological invasions on disturbance regimes. Trends in Ecology & Evolution. 1998;**13**(5):195-198. DOI: 10.1016/S0169-5347(97)01286-X

[45] Ariffin EH, Sedrati M, Akhir MF, Yaacob R, Husain ML. Open sandy beach morphology and morphodynamic as response to seasonal monsoon in Kuala Terengganu, Malaysia. Journal of Coastal Research. 2016; Special Issue 75. In: Proceedings of the 14th International Coastal Symposium, Sydney, 6-11 March 2016: pp. 1032-1036. DOI: 10.2112/SI75-207.1

[46] Penduduk Terengganu gusar hakisan pantai makin parah [Internet]. 2017. Available from: http://www.astroawani.com/berita-malaysia/penduduk-terengganu-gusar-hakisan-pantai-makin-parah-161864 [Accessed: September 3, 2018]

[47] Ir Haji Keizrul bin Abdullah. Malaysia Coastal Environment-Planning, Development and Management of Environment in Preparation for the Next Millenium. [Internet]. 2018. Available from: https://www.water.gov.my/ressources.pdf [Accessed: September 1, 2018]

[48] de Andrésa M, Barragána JM, Sanabriaa JG. Ecosystem services and urban development in coastal social-ecological systems: The Bay of Cádiz case study. Ocean and Coastal Management. 2018;**154**:155-167. DOI: 10.1016/j.ocecoaman.2018.01.011

[49] You S et al. Coastal landscape planning for improving the value of ecosystem services in coastal areas: Using system dynamics model. Environmental Pollution. 2018:1-11. DOI: 10.1016/j.envpol.2018.06.082

[50] Gunderson LH. Ecological resilience in theory and application. Annual Review of Ecology and Systematics. 2000;**31**:425-439

[51] Plieningen T, Van der Horst D, Schleyer C, Bieling C. Sustaining ecosystem services in cultural landscapes. Ecology and Society. 2014;**19**(2):59. DOI: 10.5751/ES.06159.190259

[52] Hausmann A, Slotow R, Burns JK, Di Minin E. The ecosystem service of sense and place: Benefits for human well-being and biodiversity conservation. Environmental Conservation. 2015:1-11. DOI: 10.1017/S0376892915000314

[53] Daniela TC, Muharb A, Arnbergerb A, Aznarc O, Boydd JW, Chane KMA, et al. Contributions of Cultural Services to the Ecosystem Services Agenda [Internet]. 2012. Available from: http:www.pnas.org/cgi/doi/10.1073/pnas.1114773109 [Accessed: September 10, 2018]

[54] The New Straits Times. Villages Live in Fear of Coastal Erosion [Internet]. 2017. Available from: https://www.nst.com.my/news/2017/03/213646/actionline-villagers-live-fear-coastal-erosion [Accessed: September 10, 2018]

[55] Maffi L, Woodley E. Biocultural Diversity Conservation, a Global Sourcebook. London and Washington D.C: Earthscan; 2010

[56] Jarina MJ. Diverse fishing operations of Setiu small scale fisheries. In: Faridah M, Jamilah MS, Jarina MJ, Rohani S, editors. Species, Ecosystem and Livelihood. Universiti Malaysia Terengganu; 2015. pp. 177-198

[57] Majid Cooke F. Symbolic and social dimension in the economic production

of seaweed. Asia Pacific Viewpoint. 2004;**45**(3):387-400

[58] Loures L, Loures A, Nunes J, Panagopoulos T. Landscape valuation of environmental amenities throughout the application of direct and indirect methods. Sustainability. 2015;7:794-810. DOI: 10.3390/su7010794

[59] Hamin EM, Abunnasr Y, Roman Dilthey M, Judge PK, Kenney MA, Kirshen P, et al. Pathways to coastal resiliency: The adaptive gradients framework. Sustainability. 2018;**10**:2629